The Changing Earth

The Natural Environment
Series editors: Andrew Goudie and Heather Viles

This series will provide accessible and up-to-date accounts of the physical and natural environment in the past and in the present, and of the processes that operate upon it. The authors are leading scholars and researchers in their fields.

Published

Oceanic Islands
Patrick D. Nunn

Land Degradation
Laurence A. Lewis and Douglas L. Johnson

Humid Tropical Environments
Alison J. Reading, Russell D. Thompson and Andrew C. Millington

The Changing Earth
Rates of Geomorphological Processes
Andrew Goudie

Forthcoming

Rock Slopes
Robert Allison

Caves
David Gillieson

Drainage Basin
Form, Process and Management
K. J. Gregory and D. E. Walling

Deep Sea Geomorphology
Peter Lonsdale

Holocene River Environments
Mark Macklin

Wetland Ecosystems
Edward Maltby

Arctic and Alpine Geomorphology
Lewis A. Owen, David J. Evans and Jim Hansom

Weathering
W. B. Whalley, B. J. Smith and J. P. McGreevy

The Changing Earth
Rates of Geomorphological Processes

Andrew Goudie
School of Geography,
University of Oxford

BLACKWELL
Oxford UK & Cambridge USA

First published 1995

Blackwell Publishers
108 Cowley Road
Oxford OX4 1JF
UK

238 Main Street
Cambridge, Massachusetts 02142
USA

British Library Cataloguing in Publication Data
A CIP catalogue record for this book is available from the British Library.

Library of Congress Cataloging-in-Publication Data
Goudie, Andrew.
The changing earth : notes of geomorphological processes / Andrew
S. Goudie.
p.   cm. — (The natural environment)
Includes bibliographical references and index.
ISBN 0–631–19468–1. — ISBN 0–631–19469–X (pbk.)
1. Geomorphology. I. Title. II. Series.
GB401.5.G68   1995
551.4'1—dc20   94–25818
                                                CIP

Typeset in 10 on 11½ pt Sabon by Graphicraft Typesetters Ltd., Hong Kong
Printed in Great Britain by T.J. Press Ltd, Padstow, Cornwall

This book is printed on acid-free paper

Dedicated to
Alice May Goudie

Contents

Plates

Figures

Tables

Acknowledgements

I am most grateful to Dr M. Summerfield of Edinburgh University and Dr S. Trudgill of Sheffield University, both of whom gave me most pertinent and useful comments on a first draft of this volume. I am very grateful to Lynn Murphy for typing the manuscript and to Peter Hayward and Ailsa Allen for drawing the diagrams.

The publisher and author are grateful to the following for permission to redraw and reproduce figures: Elsevier Science Ltd (figs 2.5, 3.3 and 7.4), Academic Press (figs 2.2, 2.3 and 4.7), Professor D. E. Walling (figs 2.18, 2.19, 2.20), Longman Group UK (figs 2.10, 3.1, 6.3 and 8.1) (from M. A. Summerfield, *Global Geomorphology*, figs 15.1, 11.4, 15.4 and 15.11), Scandinavian University Press (fig. 7.3), Gebrüder Borntraeger (figs 2.8, 2.9, 2.17, 3.14, 6.1 and 8.5), UNESCO (fig. 3.8), Oxford University Press (fig. 8.10), Cambridge University Press (figs 2.7 and 4.6), the *South African Geographical Journal* (fig. 5.1), the *American Journal of Science* (figs 2.12 and 2.13), the International Glaciological Society and Professor David Sugden (fig. 6.2), the International Association of Scientific Hydrology (figs 2.10, 3.6, 3.12 and 8.16), The Institute of British Geographers (figs 8.14 and 8.15), John Wiley and Sons (figs 2.6, 2.11, 2.20, 3.7, 3.9, 3.13, 3.18, 7.1 and 7.2), Harvard University Press (figs 8.2 and 8.3), Scope Publications (fig. 2.14) and Routledge (fig. 8.6). The photographs were all taken by the author.

ONE

Ideas and problems

Introduction

An understanding of rates of land surface change has always been a central focus of geomorphological research. Among the questions which require a knowledge of rates of change are the following: How important is climate as a control of geomorphological processes? To what extent are humans influencing landform development? What is the potential useful life time of a particular engineering structure? How can environments be transformed by future global environmental changes? What is the relative significance of different geomorphological processes? How much time is required for a particular landform assemblage to develop?

The study of rates of land surface change has had a long history. In the early nineteenth century geomorphologists struggled to ascertain the relative power of catastrophic and more normal and long continued events. The Uniformitarian paradigms, involving the likes of Hutton, Playfair and Lyell, were concerned with this issue, while the development of fluvialism (a belief in the power of rivers to mould the landscape) involved basic evaluations of some crucial issues, such as the time that was available for a certain degree of landscape change to occur. Arguments about the relative importance of Noah's Flood and its postulated associated 'waves of translation' (great surges and inundations of sea water) compared with long-continued fluvial denudation, and about the relative power of the sea to create planation as compared with rivers, necessitated qualitative (and sometimes quantitative) assessments of rates. By the mid-nineteenth century some discharge and sediment load data were becoming available, topographic maps were increasingly accurate and beginning to cover large parts of the Earth's surface, engineers were requiring precise data on the movement of water and sediment, and indefatigable observers were using such other sources that their ingenuity suggested were available, including the famous gravestones of St Giles' graveyard in Edinburgh (Geikie, 1880). Chorley, Dunn and Beckinsale (1964) review such developments, and point to the first quantitative revolution in geomorphology.

Moreover, as the decades passed by, so too could the length of time over which geomorphological processes were perceived to operate increase. As scientists gained an ever-increasing impression of the great antiquity of the Earth they were more able to countenance the long-term operation of 'normal' processes, and were able to depend less and less for their explanations on the operation of catastrophic processes during a very compressed period of time. Thus, whereas in the late eighteenth century the Earth was thought to be only about 6000 years old, the Creation having been calculated from biblical evidence as occurring in 4004 BC, from the 1860s William Thomson (Lord Kelvin), was advancing the idea that perhaps 100 million years had passed since the planet was a ball of molten rock. However, the discovery of radioactivity introduced an entirely new factor into calculations of the age of the Earth, and in the early 1900s scientists began to appreciate that the decay of radioactive elements in the Earth's core could provide sufficient energy to account for some billions of years of geological and geomorphological activity. At the present time, it is believed that the Earth may be around 4600 Ma old.

Some early quantifiers

An early attempt to quantify rates of geomorphological change was that by Croll (1875), who used data on the amount of material being transported down rivers for two reasons: to establish both the extent of geological time, and to show the power of subaerial forces. He suggested that in Scotland the land might be being denuded at a rate of about one foot (c.30 cm) in 6000 years or 50 mm 1000a^{-1} (p. 335):

But slow as is the rate at which the country is being denuded, yet when we take into consideration a period so enormous as 6 millions of years, we find that the results of denudation are really startling. One thousand feet of solid rock during that period would be removed from off the face of the country. But if the mean level of the country would be lowered 1000 feet in 6 millions of years, how much would our valleys and glens be deepened during that period? This is a problem well worthy of the consideration of those who treat with ridicule the idea that the general features of our country have been carved out by subaerial agency.

Croll's fellow Scot, Archibald Geikie (1868), was equally concerned to demonstrate the power of subaerial erosion in comparison with that of the sea, and provided data on suspended loads for a range of the world's rivers, expressing them as a rate of surface lowering (table 1.1). He suggested that rates of denudation were quicker than had often been supposed and averred that 'it is most unphilosophical to demand unlimited ages for similar but much less extensive denudations in the geological past' (Geikie, 1893,

Table 1.1 A. Geikie's estimates of rates of denudation for some major river basins (Geikie, 1893, p. 462)

Name of river	Area of basin in square miles	Annual discharge of sediment in cubic ft.	Fraction of foot of rock by which the area of drainage is lowered in one year
Mississippi	1,147,000	7,468,694,400	$\dfrac{1}{6000}$
Ganges (Upper)	143,000	6,368,077,440	$\dfrac{1}{823}$
Hoang Ho	700,000	17,520,000,000(?)	$\dfrac{1}{1464}$
Rhone	25,000	600,381,800	$\dfrac{1}{1528}$
Danube	234,000	1,253,738,600	$\dfrac{1}{6864}$
Po	30,000	1,510,137,000	$\dfrac{1}{729}$

p. 463). Using data for the Mississippi he estimated that North America might be worn away in four and a half million years. From the Ganges (Ganga) data he estimated that the Asiatic continent might be reduced to sea level in just 950,000 years. Using data for the Po basin he estimated that Europe might be levelled in just half a million years, whereas an analysis of British river data suggested to him that the British Isles might be levelled in about five and a half millions of years. He also suggested that rivers could excavate a 1000 foot valley (c.300 m) in about 1.2 million years (c.250 m Ma^{-1}).

By contrast Geikie felt that the power of marine denudation was considerably less (Geikie, 1893, pp. 466–7):

From the destructive effects of occasional storms an exaggerated estimate has been formed of the relative potency of marine erosion. That the amount of waste by the sea must be inconceivably less than that effected by the subaerial agents, will be evident if we consider how small is the extent of surface exposed to the power of the waves, when contrasted with that which is under the influence of atmospheric waste. In the general degradation of the land, this is an advantage in favour of the subaerial agents which would not be counterbalanced unless the rate of waste by the sea were many thousands or millions of times greater than that of rains, frosts, and streams. But in reality no such compensation exists. In order to see this, it is only necessary to place side by side measurements of the amount of work actually performed by the two classes of agents. Let us suppose, for instance, that the sea eats away a continent at the rate of ten feet in a century – an estimate which probably attributes to the waves a much higher rate of erosion than can, as the average, be claimed for them. Then a slice of about a mile in breadth will require about 52,800 years for its demolition, ten miles will be eaten away in 528,000 years, one hundred miles in 5,280,000 years. Now we have already seen that, on a moderate computation, the land loses about a foot from its general surface in 6000 years, and that, by the continuance of this rate of subaerial denudation, the continent of Europe might be worn away in about 4,000,000 years. Hence, before the sea, advancing at the rate of ten feet in a century, could pare off more than a mere marginal strip of land, between 70 and 80 miles in breadth, the whole land might be washed into the ocean by atmospheric denudation.

The cogency and the quantitative basis of Geikie's writings did much to strengthen the position of the fluvialists, and to weaken the position of the marine denudationists, of whom another Scot, Mackintosh, was perhaps the most vocal exponent (see, for example, the debate on the 'denudation controversy' in the *Geological Magazine* in 1865, and Mackintosh, 1869). Mackintosh relied more on assertion than on data and so was less persuasive, as these excerpts demonstrate:

Denuding power of Rivers not to be measured by the Quantity of Matter they carry to the Sea. – I think that a little consideration will show that all attempts to measure the denuding power of rivers by the quantity of matter they carry to the sea, must prove deceptive, unless the channels of the rivers from their mouths to their sources consist of solid rock.

As regards England and Wales, I think it may be safely asserted, that at least nine-tenths of the matter carried by many rivers into the sea is marine, or marine-glacial drift, derived from the drift in situ, or from places where it has been reassorted by the rivers. (Mackintosh, 1869, p. 275)

The quantity of rock removed by the sea greatly exceeds that removed by rains, rivers, and ice. More than two-thirds of the earth's surface are under the sea at any given time. The shape of the greater part of the bottom of the sea has, therefore, been formed by the sea. As nearly a third part of the earth's surface, constituting dry land, sinks slowly down through the zone of waves and currents (the latter ordinary or paroxysmal), which is at least 1000 feet in vertical extent, and rises slowly up through the same zone, marine agency must be sufficient to obliterate the forms of land surface left by rains, rivers, and ice; and as more than two-thirds of the whole surface are either acted on by the sea, or by the sea preserved from subaerial action, it follows that the greater part of the land surface at any given time must present the form given to it by the sea. The sea can obliterate (by denudation and deposition, if not by denudation alone) forms of land surface much quicker than subaerial agents can produce them. These agents require so comparatively long periods to give rise to new forms of ground, that the greater part of the land-surface at any given time must retain the principal features impressed upon it before it rose above the sea. The only way in which it would seem possible for the subaerialist to attempt to escape from these considerations would be, by asserting that the area of England and Wales has been quite exceptional in continuing above the surface of the sea, while numerous and extensive oscillations in level have affected the general surface of the earth. In answer to this it may be remarked that the main forms of the ground in South Britain (cwms perhaps excepted) are characteristic of every region of the globe. (Mackintosh, 1869, pp. 292–3)

The findings of Croll and Geikie were substantiated and strengthened by those of Ewing and Reade in the 1880s. Using data on water chemistry and stream discharge they pointed out the power of chemical denudation (a topic largely ignored by Geikie, who was in essence concerned with the movement of suspended sediment in rivers), thereby adding yet more power to those who championed the role of rivers in landform development (the fluvialists). Ewing (1885) calculated a rate of surface lowering by solutional denudation for a limestone area of the Appalachians (1 m per 29,173 years, or $c.34$ mm 1000 a^{-1}), thereby establishing a method that was to be taken up in the 1950s by Corbel (1959), while Reade (1885) looked more generally at the quantity of dissolved material in rivers being evacuated from the Americas and elsewhere. For the Mississippi he calculated that if one solely employed the suspended load then the rate of surface lowering was 1/6000th of a foot per year ($c.51$ mm a^{-1}), whereas if the chemical load was added as well the rate increased to 1/4500th of a foot per year (67 mm a^{-1}). On a global scale he estimated the approximate balance of solutional and suspended load (p. 298): 'It would appear from the examples of the Mississippi, the Nile and the Danube that the matter brought down in solution and suspension is as 1 to 3.'

Developments in the twentieth century

However, the Davisian geomorphology that followed was less concerned with either processes or rates, and much attention was directed in the first half of the twentieth century to the study of landform evolution and history. There were, or course, notable exceptions, and the literature of the Dust Bowl and the Tennessee Valley Scheme, for instance, is replete with detailed studies of rates of soil erosion (see, for example, Bennett, 1939). Indeed, such was the fear that recent agricultural expansion in the USA was causing dreadful devastation (see McGee, 1911; and Shaler, 1912) that erosion plot studies were started in Utah in 1912 and at the Missouri Agricultural Experiment Station in 1917. From 1929 onwards, with L. A. Jones, H. H. Bennett supervised the establishment of ten further erosion experiment stations in Oklahoma, Texas (2), Kansas, Missouri, North Carolina, Washington, Iowa, Wisconsin and Ohio (Meyer, 1982).

Another notable exception was in Great Britain, where the Royal Commission on Coastal Erosion (1911) set out in a somewhat rambling way 'to reach some conclusions with regard to the amount of land which has been lost in recent years by the encroachment of the sea on the coasts of the United Kingdom and to the amount which had been gained by reclamation or accretion of the sea'. It used the evidence from Ordnance Survey maps, together with information provided by local authorities and coastal landowners.

However, it was not until a third of a century ago that there was to be a truly concerted attack on the question of determining the rate of operation of a wide range of geomorphological processes. The situation has been summarized thus by Saunders and Young (1983, p. 473):

In 1960 little was known about the absolute rates of operation of geomorphological processes on slopes. Reasons were the slowness of such activity in comparison with a human lifetime, the previous concentrations of research effort on landforms rather than processes, and the low level of technical sophistication of most geomorphologists at the time.

A revolution in geomorphology followed, which led to as much or more attention being paid to processes as to form. . . . A steady stream of publications emerged reporting rates of processes, particularly surface processes on slopes and rates of total denudation . . . trying to keep up with current publication is like cleansing the Augean stables.

A concern with rates of operation of geomorphological processes arose from the mid-1950s in a variety of countries with a range of different traditions. For example, in Poland geomorphologists undertook distinguished and detailed evaluations of processes operating in polar regions and periglacial environments (e.g. Jahn, 1961), and also developed techniques for assessing the intensity of soil erosion processes (e.g. Reniger, 1955). Scandinavian geomorphologists were also active at much the same time and the comprehensive studies at Karkevagge (near the Arctic Circle) by Rapp (1960) remain a model for this type of work, not least because he was able to discredit so many armchair-based shibboleths relating to the speed of process operation in cold environments. Of no lesser significance was the work by a frenchman, Corbel, and his co-workers (Corbel, 1959) on rates of chemical denudation in many different parts of the world. There were also various teams in the USA, especially those concerned with fluvial processes (e.g. Leopold, Wolman and Miller 1964), who started to think deeply about the question of frequency and magnitude of formative geomorphological events. Likewise, one of the first symposia of the newly established British Geomorphological Research Group, in 1965, was on rates of denudation even though many of its first officers appeared to be staunch denudation chronologists.

One important development in such studies was the desire to produce sediment budgets for drainage basins. These are a quantitative accounting of sediment entering, leaving and being stored within a basin. As Abrahams and Marston (1993, p. 221) have put it:

A sediment budget, however, is more than just a mass balance equation for a basin. It identifies the components of the drainage basin erosion-transport-deposition system, measures the processes operating in these components, and quantifies the relations between the components. As such, the sediment budget provides a framework for investigating the movement of sediment through a drainage basin and interpreting landform change.

The production of sediment budgets is a crucial means for under-standing geomorphological systems and has been the focus of considerable work from the 1980s onwards.

Although many of these studies developed because of a dis-satisfaction with some aspects of Davisian geomorphology, and because of a basic fascination with the evolution and functioning of geomorphological systems, there is no doubt that some of the increasing body of work to determine rates of geomorphological change was stimulated because of their perceived relevance in the field of engineering and landscape management. As Thornes and Brunsden (1977, p. 96) point out:

One need not seek far for examples; reservoir siltation, beach aggradation, cliff recession and soil erosion are crucial problems and study of rates, as well as the controls of such rates, is essential. There is, however, a certain irony, for while the pragmatic demands on the data on rates are such as to involve high penalties for providing the wrong rate, there is usually a sense of urgency that precludes long-term investigation.

The problems

The pressure to use short-term data, to which Thornes and Brunsden have just referred, is but one of many general problems that attend attempts to estimate rates of geomorphological change.

A first, but still central, problem which one faces if one is to comprehend and compare different estimates of rates of change is that they are often reported in different forms or units. Thus many of the classic studies of rates of denudation employ data derived from discharge of material by streams and consequently are ex-pressed as the mass or volume of sediment derived from a catch-ment (e.g. in kg, t, m^3 or ppm of material) averaged for a unit area of the catchment in a unit period of time to give a mean rate of land surface lowering. By contrast, data on transport rates of material on hill slopes are usually defined by the velocity at which the material moves (e.g. in metres per year or m a^{-1}) or by the discharge of the material through a contour length (e.g. m^3 m^{-1} a^{-1}). Equally, because of problems of determining appropriate den-sity values for different types of soil, sediment and rock there can be difficulties with converting mass per unit area to volume per unit area and vice versa.

One attempt to develop a measure of erosional activity that is applicable to hill-slope and river channel processes alike is that of Caine (1976). He proposed that erosion involves sediment move-ment from higher to lower elevations and so is a form of physical work. It requires a reduction in the potential energy of the land-scape in which there are successive conversions from potential to kinetic energy that is finally dissipated as heat and noise. The change in potential energy through time is defined as

$$\Delta E = mg(h_1 - h_2)$$

in which m is the mass of sediment involved in the movement, g is the gravitational acceleration (9.8 m s⁻¹) and h is elevation, with $h_1 - h_2$ being the change in elevation during the interval between time 1 and time 2. The dimensions of work (E) are joules. For most stream channel situations:

$$\Delta E = vpg(h_1) \quad \Delta E = vpg(h_1 - h_2) \times 10^3$$

where v is the volume of sediment (in cubic m³) and p is its density (in t m⁻³), and for slope measurements across a unit contour length when the thickness of the moving mantle is known directly:

$$\Delta E = vpg(d \sin \theta)$$

where d is the slope distance (m) and θ is the slope angle.

In terms of comparability of units, a minority of workers employ the Bubnoff Unit, especially for slow processes characteristic of tectonic and denudation change. The Unit (B) is as follows:

$$1B = 1 \text{ m per million years}$$
$$1B = 1 \text{ mm per 1000 years}$$
$$1B = 1 \text{ }\mu\text{m per year}$$

For the sake of comparison I have, as far as possible, used SI and SI-related units in this book and have converted as many figures as possible either into mm a⁻¹ or into t km⁻² a⁻¹. Where appropriate or where conversion is problematical, however, I have sometimes kept the original units used by individual geomorphologists.

A second difficulty with evaluation of all rates of geomorphological change is that there has been an understandable tendency for workers to undertake investigations in dynamic areas, or to investigate a particular process in a particular location because it is plainly highly effective and important. There is a bias against making measurements in situations where little activity is apparent.

A third difficulty is one of scale. Rates determined at one spatial scale cannot necessarily be extrapolated to another spatial scale. It is, for instance, difficult to scale up from erosion plot studies to an entire catchment, or to scale up from small catchments to large. Part of the problem is accounted for by storage within the system (Meade, 1982). For example, in many cases slope debris may be translocated within a basin but not exported from it, being stored as colluvium in screes and fans or on terraces and floodplains. A scale effect may operate here, for hill-slope debris is commonly delivered directly into the channel in a low order stream, but is held in store in higher order catchments. Thus, in a study of river basins in the south-eastern USA Trimble (1977) found that while upland erosion was progressing at about 9.5 mm 1000 a⁻¹, sediment yields at the mouths of catchments were only 0.53 mm 1000 a⁻¹. The delivery ratio was thus only 6 per cent, with the difference being stored in valleys and channels.

Numerous factors may control the proportion of eroded sediment that is actually exported from a catchment; they include the type, magnitude and proximity of the sediment source to the drainage network, the grain size of the sediment, the availability of sediment storage areas, and catchment characteristics such as mean slope. Basin area (A) may also affect the proportion of material that is evacuated from the system, for there is a relationship

$$Y \propto A^b$$

where the exponent b is negative, and is approximately -0.12, but approaches zero for large drainage basins (Glymph, 1951). This is because topographic factors are related to basin area. Thus different values for rates of erosion per unit area are often obtained depending upon whether the study area is large or small. This is, among other things, due to the different transport lengths involved. Jansson (1982, p. ii) has expressed this well:

Plot studies give a measure of the effect of splash, sheet and rill erosion and the transport of the eroded material to the end of the plot. Sediment yield from small drainage basins implies that the material has been transported over greater distances, first in interfluvial areas to a small channel and then through the drainage net to the point of measurement. Material from gullies, channels and mass movements is included. Within drainage basins there is always some deposition at the end of slopes, in depressions, in lakes and reservoirs, on floodplains and in channels. These conditions are still more pronounced in large drainage basins. Therefore, it is impossible to compare directly rates of erosion for different regions and to use plot values in one case and values from drainage basins in another.

Indeed, table 1.2, which is derived from numerous sources in Jansson (1982, table 5), suggests that while there is a considerable scatter in the correlations between sediment delivery and basin area, the sediment yield varies with the -0.04 to -0.47 power of the drainage basin area. This inverse relationship can be surmised to be due to several factors. First, small basins may be totally covered by a high intensity storm event, giving high maximum erosion rates per unit area. Secondly, the relative lack of floodplains in small basins gives a shorter sediment residence time in the basin and a more efficient sediment evacuation. Third, small basins often have steeper valley side slope angles and stream-channel gradients, encouraging more rapid rates of erosion.

Within a catchment there will be marked local spatial variation in the operation of individual processes, so that a detailed sampling programme may be required if some reasonably representative value is to be obtained. As Carson (1967) demonstrated, some soil properties, including infiltration capacity, have a tendency to show a large range over small areas, and this variability will inevitably be reflected in the operation of processes related to it, such as overland flow. Equally, different parts of a drainage basin contribute sediment at very different rates. Building on the work of Gibbs (1967) Chorley et al. (1984, p. 56) remark:

Table 1.2 Relations between sediment yield and catchment area (SAY = sediment yield (m^3 km^{-2} a^{-1}), Y = sediment yield (t km^{-2} a^{-1}), X = area (km^2))

Source	Position	Basin size (km^2)	No. of basins	Equation
Scott et al. (1968)	Southern California	8–1036	8	$SAY = 1801\ X^{-0.215}$
Fleming (1969)	Mostly American, African and UK basins		235	$Y = 140\ X^{-0.0424}$
Sediment. Eng. (1975)	Upper Mississippi River; a varies with 'land resource areas'			$SAY = aX^{-0.125}$
Strand (1975)	South-western USA		8	$SAY = 1421\ X^{-0.229}$
	Equations for the upper limits of the silting rate in reservoirs			
Khosla (1953)	Different parts of the world	< 2509	89	$SAY = 3225\ X^{-0.28}$
Joglekar (1960)	95 American, 24 Indian 10 European, 5 Australian and 5 African basins		139	$SAY = 5982\ X^{-0.24}$
Varshney (1970)	Northern Indian mountain basins (north of Vindhyas)	< 130		$SAY = 3950\ X^{-0.311}$
	Northern Indian plain	< 130		$SAY = 3920\ X^{-0.202}$
	Northern India	> 130		$SAY = 15340\ X^{-0.264}$
	Southern India	< 130		$SAY = 4600\ X^{-0.468}$
	Southern India	> 130		$SAY = 2770\ X^{-0.194}$

Source: from various authorities cited in Jansson (1982, table 5).

Ninety-five per cent of the load of the Mississippi River basin is derived from 5 per cent of its headwater area, and 82 per cent of the load of the Amazon River is provided by the 12 per cent of the basin occupied by the Andes. Even for small basins, most of the surface sediment is derived from the 'partial areas' of low infiltration capacity, which for areas in the north-eastern United States occupy an average of 15 per cent of the basins in the summer and 50 per cent in the spring.

As well as being highly variable according to spatial scale and location, the rate of operation of geomorphological processes may be highly variable in time as a response to miscellaneous environmental changes and fluctuations. As Selby (1985, p. 576) has remarked, 'In many discussions of the rates of the landform change emphasis is placed upon "average rates" of denudation, uplift, sea-floor spreading, and other processes. Such figures are valuable because they can give an impression of the rate of long-term change of the Earth's surface. Average rates, however, can also obscure recognition of the fact that many natural processes are episodic in operation, and also that land surfaces may evolve in a series of leaps, with periods of stability followed by brief periods of severe erosion.' Comprehension of the importance of frequency and magnitude of process is therefore fundamental (Wolman and Miller, 1960). The phenomena that are responsible for erosion and landscape formation range from those that are relatively uniform or continuous in operation (e.g. removal of matter in solution by groundwater flow), through seasonal, periodic and sporadic phenomena to catastrophic phenomena that are abnormal and anomalous (e.g. a major flood caused by an extreme combination of meteorological and hydrological conditions). In some environments the frequent events of moderate magnitude do the greatest proportion of work, in others the 'continuous' phenomena such as solution may be dominant, while in others catastrophic mass movement processes may dominate. It is clear that the periodicity and frequency of dominant and formative events in any given area may vary widely, depending on the processes involved and the variables that control thresholds of activity and recovery rates. Thus the reliability and utility of short-term monitoring experiments of rates may be questionable, and there may be some advantages to be gained from using longer-term historical, archaeological and sedimentological sources (e.g. Hooke and Kain, 1982).

In any case, where process monitoring studies have been continued for more than the three years of a Ph.D. investigation, discrepancies have often been encountered between values obtained for just a few years of study and those obtained for a decade or more. This may be a consequence of such factors as initial site disturbance causing atypical conditions in the early years (see, for example, Trudgill et al., 1994).

In addition there are the considerations posed by longer term changes in climate at various time-scales, such as the Little Ice Age, the Medieval Little Optimum, the Holocene neoglaciations, the early Holocene pluvial of low latitudes, the major glacials, interglacials, stadials and interstadials of the Pleistocene, not to speak of all the events of pre-Pleistocene times. Palaeoclimatology teaches us that climatic and other environmental changes have often been frequent, rapid, and of a marked degree. Thus we cannot necessarily accept present-day measurements, which are in an environment that is probably atypical of the Pleistocene and even more atypical of larger time-scales, as being representative of geological rates of change (figure 1.1).

The episodic nature of much change is another major concern that can be demonstrated through the study of rates of sediment accumulation. Sadler's (1981) examination of nearly 25,000 rates of sediment accumulation shows that they span at least 11 orders of magnitude. More importantly, he also found that there is a systematic trend of falling mean rate with increasing time span. Although measurement error and sediment compaction contribute to such trends, they are primarily the consequence of unsteady, discontinuous sedimentation. Thus, once again, short-term rates of accumulation are a very poor guide to long-term accumulation regimes.

Sadler's conclusions were very much supported in the context of more general geomorphological and tectonic processes by Gardner et al. (1987, p. 259):

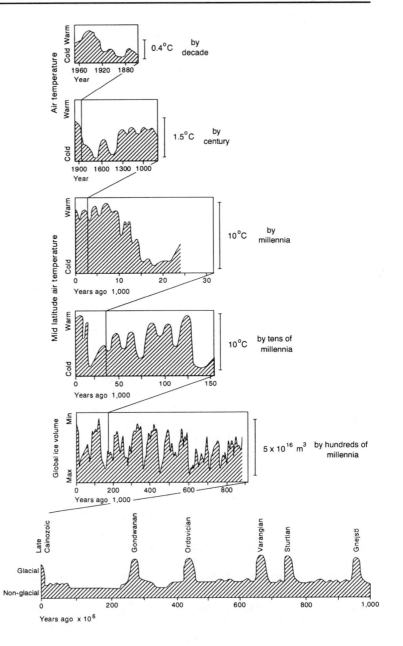

Figure 1.1
The different scales of
climatic change over the
past billion years (source:
Goudie, 1993, fig. 1.3).

Where plots of rate measured over different time intervals show systematic changes with age (absolute time), it is often inferred that the process rate has changed with time. The implicit assumption is that dividing observed change by elapsed time removes the effect of time; that is, the rates are independent of the measurement interval. Plots of this type have led researchers to attribute such changes in process rate to temporal variation in tectonic activity, climatic change, complex response of the geomorphic system, or human activity, among others. . . . Inferences of this type may be true in some cases, but . . . raw data estimated from different time intervals do not provide the sufficient condition because process rates are, in fact, a function of the measured time interval.

A further major problem with trying to extrapolate present-day estimates of geomorphological change back through time is that 'natural' rates have been modified to an increasing degree over the last two or three million years by the presence of humans in the landscape. The use of fire (for the last 1.4 million years), of deforestation (for the last 9000 years), of ploughing (for the last 5000 years), grazing, excavation, construction and pollution are among the actions that have sometimes dramatically altered the rates of operation of geomorphological processes. The topic is so vast (see Goudie, 1993, for a review) that it is only possible to give a few examples in this introduction to indicate the magnitude of the changes involved. Many other examples will be given elsewhere in this volume.

Table 1.3a displays some data for south-western Australia on the changes in the chloride loadings of streams that have been cleared of native Eucalyptus forest for agricultural purposes. Chloride loss in streamflow for forested areas averages 14.2 t km^{-2} a^{-1}, whereas for the farmed area the figure has gone up by over three-fold to 47.6 t km^{-2} a^{-1}. Table 1.3b displays comparable data on nitrate losses following clear-cutting of forests in the USA. The mean loss for forested sites is 0.144 t km^{-2} a^{-1}, while that for disturbed catchments is 2.68 t km^{-2} a^{-1}, representing an almost nineteen-fold increase. Table 1.3c displays some data for rates of erosion caused by urbanization and construction in the USA. Normal background rates of erosion in the forest are of the order of 10^1 t km^{-2} a^{-1}, on agricultural land of the order of 10^2 to 10^3 t km^{-2} a^{-1}, while during the construction phase they are more likely to be of the order of 10^2 to 10^4 t km^{-2} a^{-1}.

By contrast, the construction of dam and reservoirs across rivers has tended in some cases to greatly reduce the discharge of suspended sediment to the oceans. Figure 1.2 shows the decline in sediment passing along the Missouri and Mississippi Rivers over the period 1939 to 1982. Downstream sediment loads have been reduced by about half during this period.

Even some great 'natural' wonders like the Niagara Falls of North America are now in reality hugely modified as a result of anthropogenic change. Not only has the amount of flow passing over the falls been substantially reduced in comparison with background levels, but so, as a consequence have the recession rates of the falls been considerably lessened (see table 1.4).

For all these sorts of reasons backward extrapolation of currently observed rates must only be undertaken with extreme circumspection. The justification for this postulate has been well expressed by Thornes and Brunsden (1977, p. 103):

Table 1.3 Selected examples of the power of the human impact on geomorphological processes

(a) Hydrological and chemical data for south-western Australia illustrating the effects of deforestation on groundwater recharge and river water chemistry

Catchment	Chloride input in precipitation ($t\ km^{-2}\ a^{-1}$)	Chloride loss in streamflow ($t\ km^{-2}\ a^{-1}$)	Recharge before clearing ($mm\ a^{-1}$)	Recharge with farming ($mm\ a^{-1}$)
(A) FORESTED				
Julimar	5.3	7.8	–	–
Seldom seen	12.0	16.0	–	–
More seldom seen	12.0	14.0	–	–
Waterfall gully	11.0	18.0	–	–
North Danalup	13.0	18.0	–	–
Davies	13.0	14.0	–	–
Yarragil	9.7	13.0	–	–
Harris	8.4	13.0	–	–
(B) FARMLAND				
Brockman	8.0	34.0	8	73
Wooroloo	7.8	42.0	4	61
Dale	2.4	46.0	0.8	24
Hotham	4.8	37.0	2	26
Williams	3.1	65.0	1	37
Collie East	5.0	74.0	2	60
Brunswick	11.0	35.0	70	500

(b) Nitrate-nitrogen losses from control and disturbed forest ecosystems

Site	Nature of disturbance	Nitrate-nitrogen loss ($t\ km^{-2}\ a^{-1}$) Control	Disturbed
Hubbard Brook	Clear-cutting without vegetation removal, herbicide inhibition of re-growth	0.2	9.7
Gale River (New Hampshire)	Commercial clear-cutting	0.2	3.8
Fernow (West Virginia)	Commercial clear-cutting	0.06	0.30
Coweeta (North Carolina)	Complex	0.005	0.73*
H. J. Andrew's Forest (Oregon)	Clear-cutting with slash burning	0.008	0.026
Alsea River (Oregon)	Clear-cutting with slash burning	0.39	1.54
MEAN		0.14	2.68

* This value represents the second year of recovery after a long-term disturbance. All of the other results for disturbed ecosystems reflect the first year after disturbance.

(c) Rates of erosion associated with construction and urbanization in the USA

Location	Land use	Rate ($t\ km^{-2}\ a^{-1}$)
Maryland	Forest	39
	Agriculture	116–309
	Construction	38,610
	Urbanized	19–39
Maryland	Rural	22
	Construction	337
	Urban	37
Maryland	Forest and grassland	7–45
	Cultivated land	150–960
	Construction	1600–22,400
	Urban	830
Virginia	Forest	9
	Grassland	94
	Cultivation	1876
	Construction	18,764
Detroit	General non-urban	642
	Construction	17,000
	Urban	741
Wisconsin	Agricultural	< 1
	Construction	19.2

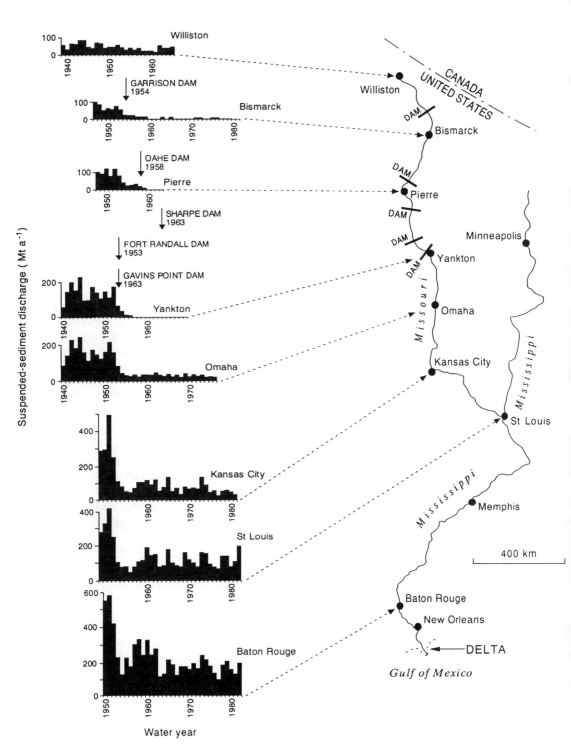

Figure 1.2
Suspended sediment discharge on the Mississippi and Missouri Rivers between 1939 and 1982 (source: Meade and Parker, 1985, with modifications).

Table 1.4 Recession rates and flow abstraction for the Niagara Falls

Period	% of total flow after water abstraction for power generation	recession rate (m a⁻¹)
1842–1905	100	1.28–1.52
1905/6–1927	72	0.98
1927–1950	60	0.67
1950–present	34	0.10

Source: K. J. Tinkler, (1993).

In *all* rates of operation in fluvial processes, we must seriously consider the utility of contemporary measurements in the role of longer term evolution of the landscape. The dictum that the present is the key to the past is perhaps more doubtful and yet more critical than in any other area of geomorphology. Insofar as a knowledge of rates is essential to the study of mechanisms in the contemporary environment, there can be little criticism. The main objection is the extrapolation of arguments about the relative importance of different rates. This is because (i) the past environments are known to have been highly variable in time and space, (ii) present estimates of many rates are by and large poor in quality and have high standard errors, (iii) we have little evidence to support the assumption that such rates are the direct response to immediate processes, given the possible transitory nature of the system state because of (iv) the widespread, almost ubiquitous impact of human activities.

Thus there are some daunting problems in assessing and using the great mass of empirical data that exists in the literature and which has accumulated over the last two centuries. There are also many processes and many environments for which our information is still highly imperfect or selective. Nonetheless, in the following chapters the attempt will be made to assemble and assess what is known about rates of change for major groups of processes.

TWO

Weathering and the dissolved loads of rivers

Introduction

Weathering is one of the most important geomorphological processes, occurring when the rocks and sediments in the top metres of the earth's crust are exposed to physical, chemical and biological conditions much different from those prevailing at the time the rocks were formed. In general two main types of weathering are recognized. Mechanical or physical weathering involves the breakdown or disintegration of rock without any substantial degree of chemical change taking place in the minerals that make up the rock mass. It incorporates such processes as frost and salt weathering and may also be achieved by organic or biological means (e.g. root wedging). Chemical weathering, in which biological processes may play a major role, involves the decomposition or decay of rock minerals through such processes as hydration, hydrolysis, oxidation and reduction, carbonation and chelation. In most parts of the world both types of weathering may operate together though in differing proportions, and one may accelerate the other. For example, the physical disintegration of a rock will expose a greatly increased surface area to chemical attack.

Techniques to estimate rates of weathering

One of the first methods that was used to estimate rates of weathering was to measure the decay of tombstones of known age (see Goudie, 1990). Geikie (1882), for example, investigated 'the older burial grounds in Edinburgh', though he appreciated that rates obtained in such an environment might be exaggerated for 'there must be a good deal more of such chemical change where sulphuric acid is copiously evolved into the atmosphere from thousands of chimneys than in the pure air of country districts' (p. 184). This is a method that has subsequently been developed by Goodchild (1890), Matthias (1967), Rahns (1971), Winkler (1975) and Reddy

Table 2.1   Methods of determining rates of weathering

1. *Hydrological observations*
   Discharge and water quality analyses
2. *Instrumental direct observations*
   Micro-erosion meters
   Plaster casts
   Rock tablets
   Carousels
3. *Archaeological*, through the degree of weathering on:
   Tombstones
   Buildings
   Barrows
   Shell middens
   Reclaimed land
4. *Geomorphological*, through the degree of weathering on phenomena of
   known age:
   Drained lake floors
   Moraines
   Dunes
   Volcanic eruptions
   Erratics on limestone pavements
   Striated limestone pavements or upstanding quartz grains
   Terraces
5. *Oceanic*
   Isotopic composition of sea-water

(1988). Because of the excellence of dating control it also permits workers to assess how rates have changed through time in response, for example, to changing atmospheric quality in cities (Cooke et al., in press). Dated buildings and other structures (e.g. sea walls) provide another opportunity for study of weathering rates and of lithological effects (e.g. Hirschwald, 1908; Barton 1916; Emery, 1960). Such work has been especially important in drawing attention to the great rapidity with which salt weathering processes can cause building disintegration (see Goudie, 1985 for a review), and to the acceleration of weathering rates in polluted urban environments (see Trudgill et al., 1991).

No less productive as a method has been the examination of weathering development (e.g. rind or varnish formation) on geomorphological surfaces of known age (e.g. moraines, glaciated surfaces, sand dunes, drained polders and lake floors, and the products of volcanic eruptions) (table 2.1).

Although determination of rates of weathering on tombstones, man-made structures and geomorphological surfaces of known age may at first sight appear somewhat crude, it needs to be appreciated that such methods give a representation of the rate of weathering over a much longer time scale than is provided by studies based upon various types of *in situ* monitoring procedure or on drainage water analyses. Indeed, there are very few methods available for establishing rates of chemical denudation over very long time spans (i.e. millions of years). One promising exception, though still problematic and controversial (Berner and Rye, 1992; François and Walker, 1992; Edmond 1992), is the use of the strontium isotopic composition of seawater through time, as determined from samples of planktonic foraminifera in ocean cores. Hodell et al. (1990) believe that changes in the quantity and proportion of strontium isotopes in seawater would be caused by changes in the riverine fluxes of strontium, which in turn would be related to changes in rates of chemical weathering on the continents. They found a generally upward trend over the last 8 million years, with especially rapid rises between 6 and 4 million years ago, and since 2.5 million years ago until the present. They attribute the increase to accelerated glaciation, uplift, the exposure of continental shelves due to lowered sea levels, and to exhumation of strontium rich Precambrian shield areas by glacial erosion. Likewise Clemens et al. (1993) found that the technique gave some indication of lower rates of chemical weathering in arid glacials than in more humid interglacials.

There are two main types of *in situ* monitoring using direct instrumentation: the micro-erosion meter and rock tablets. Both methods have predominantly been employed in limestone terrains, but there is no reason, given a long enough sampling period, why they should not be employed on other rock types, as was done by Day et al. (1980) using Silurian mudstone. Details of the micro-erosion meter technique and its advantages and problems are provided by High and Hanna (1970), Trudgill et al. (1981) and by Viles and Trudgill (1984), while a discussion of the rock tablet technique occurs in Trudgill (1975 and 1977). Crowther (1983) evaluates the rock tablet technique against techniques based on water analyses and finds major discrepancies, while Jaynes and Cooke (1987) have used rock tablets attached to a rotating carousel to determine rates of rock weathering in urban atmospheres. Caine (1979) used crushed material rather than tablets, and suggested that as it exposes a greater surface area to weathering processes it may be more effective at detecting modest rates of change in a finite period.

The final field method of determining rates of weathering is to determine the amounts of dissolved material being transported in river water. A detailed examination of this important and much used method will form the second part of this chapter.

Duricrust formation

Especially in low latitudes, many land surfaces are mantled by various types of duricrust of which calcretes, silcretes, ferricretes (laterites) (plate 2.1) and alcretes (bauxites) are the most prevalent. Such deposits, which can attain considerable thicknesses, can be defined (Goudie, 1973, p. 5) as 'a product of terrestrial processes within the zone of weathering in which either iron and

Plate 2.1
A laterite duricrust at Panchgani in Maharashtra, western India, developed on the passive margin of the Indian Plate. Thick duricrusts and associated deep weathering profiles probably take a long time to form.

aluminium sesquioxides (in the case of ferricretes and alcretes) or silica (in the case of silcrete) or calcium carbonate (in the case of calcrete) or other compounds in the case of magnesicrete and the like have dominantly accumulated in and/or replaced a pre-existing soil, rock or weathered material, to give a substance which may ultimately develop into an indurated mass'.

The various duricrust types have a series of important geomorphological effects. These effects result from the thickness of the profiles, the properties of the different components of the profiles (e.g. their occasional ability to harden on exposure), their tendency to form caprocks, and the topographic situations in which the duricrusts develop.

It is difficult to give any precise figures for rates of duricrust formation, though there are some well-documented records of calcrete crusts having formed in living memory or having affected

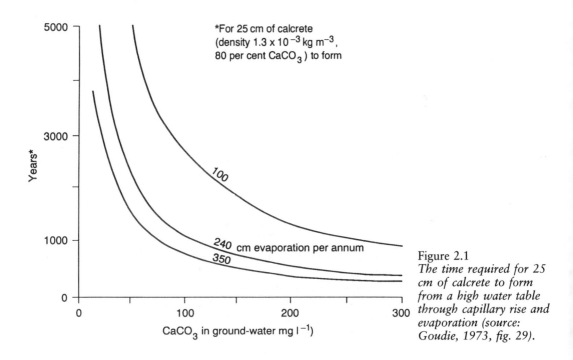

Figure 2.1
The time required for 25 cm of calcrete to form from a high water table through capillary rise and evaporation (source: Goudie, 1973, fig. 29).

datable archaeological material. In north Africa, for example, crusts have formed over second world war graves near Algiers (Pouquet, 1966), within 75 years in the Chelif Valley of central Algeria (Boulaine, 1958), over Roman remains at Volubilis in northern Morocco (Dalloni, 1951), and since the start of cereal cultivation in French colonial times in the Algerian high plateaux (Charles, 1949).

Some approximation of the time-scales involved in duricrust formation can be gained by various empirical formulae. In the case of calcretes formed by evaporation from a shallow water-table

$$T = \frac{C_t C_d CaCO_3\%}{S_{CaCO_3}E} \times 10^4$$

where T is the time in years, C_t is the thickness of the calcrete profile in cm, C_d is the dry density of the calcrete in kg m^{-3}, $CaCO_3$ is the carbonate content of the profile in per cent, S_{CaCO_3} the calcium carbonate content in ppm of soil water, and E is the evaporation in cm per year. Figure 2.1, which is based on this formula, indicates the time required to form a 25 cm layer of calcrete hardpan (density 1.3×10^{-3} kg m^{-3}, and 80 per cent $CaCO_3$) in areas with evapotranspiration ranging from 100–350 cm a^{-1}, and soil water calcium carbonate contents of 0–300 mg l^{-1}. Under such conditions a 25 cm calcrete might form in just a few thousand years.

A similar formula for use with calcretes formed by downward leaching of dust-derived carbonates (the *per descensum* model) is:

$$C_t = \frac{ID_{CaCO_3}}{C_g C_{CaCO_3}}$$

where C_t is the thickness of calcrete formed in 10,000 years, I is the input of dust in g m^{-2} a^{-1}, D_{CaCO_3} is the calcium cabonate content of the dust, C_g is the specific gravity of the calcrete, and C_{CaCO_3} is the calcium carbonate content of the calcrete. It is assumed that no carbonate is completely leached from the system. Using data on Texan dust deposition inputs this formula becomes:

$$C_t = \frac{24 \times 25}{2.25 \times 80} = 3.33 \text{ cm per } 10,000 \text{ years}$$

This is equivalent to about 1 m in 300,000 years, a rather low value.

An alternative formula is developed from an idea of Gardner (1968) who proposed that an approximate age of geomorphic surfaces in Nevada, USA, could be obtained by estimating the quantity of carbonate-saturated water required to deposit a given volume of carbonate:

$$W = \frac{C_{CaCO_3} - S_{CaCO_3}}{4 \times 10^{-2} \text{ g } l^{-1}}$$

where $C_{CaCO_3} - S_{CaCO_3}$ is the difference in weight of limestone between the calcrete and the parent material (i.e. the weight of secondary carbonate) for a given thickness of profile with a 1 cm^2 cross section, and 4×10^{-2} g l^{-1} is the theoretical solubility of calcite at 25°C and at a CO_2 pressure of $10^{-3.5}$. Gardner found that a 124 cm calcrete layer with a density of 2.62×10^{-3} kg m^{-3} and an insoluble residue of 17.2 per cent underlain by a same thickness of transitional material would develop in 890,000 years with 115 mm mean annual rainfall and in 440,000 years with 230 mm rainfall. This is equivalent to one metre of calcrete hardpan in 720,000 to 360,000 years.

For laterites, Trendall (1962) made the assumption that 20 units of bedrock, when fully weathered, would produce approximately one unit of iron-rich laterite. He therefore estimated that in the case of a 30-foot thick Uganda laterite it was the residue of no less than 600 feet of weathered bedrock. Using a figure for the silica content of bedrock and drainage waters, and making an assumption about the amount of water that percolated through the system each year, he calculated that the laterite must have taken a great deal of time to form: 23.5 million years (c.1 m in 2.35 million years). A similar method was used by Cooper (1936) in Ghana.

He calculated that it would take at least 7 million years to form a 60-foot thick bauxite (a rate of around one-third of a million years per metre).

The amount of time required for a duricrust to form depends partly on the nature of the sediment or rock being affected. Alexander and Cady (1962), for example, have pointed out that iron accumulation is required in greater quantities to give a hard ferricrete when the host material is one with a fine texture. Likewise, in the case of calcretes, Flach et al. (1969) maintain that because of the relatively small amount of pore space, continuously plugged and cemented horizons form much more rapidly in very gravelly soils than in non-gravelly soils.

In general, however, it seems likely that the quickest rates of durictust formation take place where there is absolute accumulation rather than relative accumulation of the key elements. Obviously, formation of a crust from solid, unweathered bedrock can be no faster than the time to reach a high degree of weathering. Because of its relatively modest solubility under most pH conditions, for example, it takes a long time for silica to be sufficiently removed in solution for relative accumulation of residue to occur. By contrast, in materials subject to enrichment (absolute accumulation) from outside sources (e.g. dust, lateral water seepage, or evaporation of groundwater *per ascensum*) the rate of development may be considerable, though even here the low equilibrium solubility of silica implies that formation of silcrete (through absolute accumulation) is also slow.

One aspect of duricrusts may be the speed with which they harden on exposure. This has both geomorphological and engineering significance. There are certainly many eye-witness reports of rapid laterite induration taking place. Harrison (1911) found that a laterite boulder hardened within a few days when removed from the soil; Sherman et al. (1953) noted the speedy development of a hardened crust on new irrigation ditch walls in Hawaii; and Alexander and Cady (1962) and Sombroek (1966), working in Guinea and Amazonia respectively, found that road cuts exposed to the sun hardened to a crust in a few years. In Cameroon, Persons (1970) estimated that following German timbering operations in the late nineteenth century, just under two metres of induration could take place in approximately a century.

The fact that many deep weathering profiles appear to be old, in that they may be located on old erosion surfaces in continental interiors, does not necessarily mean that the saprolite of the deep weathering profiles formed slowly. Moreover, rates of ferricrete or bauxite formation may be slower than the rates for less completely altered weathering profiles. Indeed, it is very likely that where there is active deep water flow caused by uplift and fracturing of rocks on upwarped passive continental margins, very rapid decay may occur (plate 2.2). In such circumstances, according to Thomas (1992), deep weathering profiles may have formed at such rates

Plate 2.2
*A deeply weathered profile developed on granite at Chapman's Peak in the Cape Province of South
Africa. On uplifted continental passive margins such deep profiles may form more quickly than has
often been maintained.*

(he cites figures as high as 60 to 150 metres in a million years) that
they may be of Quaternary age.

The formation of rock varnish and rinds

Rock varnish and rock rinds are, like their thick cousins, duricrusts,
the products of chemical change. Rock varnish, sometimes also
called 'desert varnish' (Whalley, 1983, p. 197) can be defined as
'a surface stain or crust of manganese or iron oxide'. It tends to
be thin, and is extensively developed in dry lands. It can be distinguished from rock rinds. These are 'oxidation phenomena which
stain the parent rock red-yellow when exposed to air or near-
surface groundwater for some time. Rinds may extend more than
a millimetre into the rock, while desert varnish is generally much
thinner.' Rinds tend to occur in more humid areas than desert
varnish.

Rock varnish (plate 2.3) has been the subject of a voluminous
and sometimes contradictory literature, for there are various theories

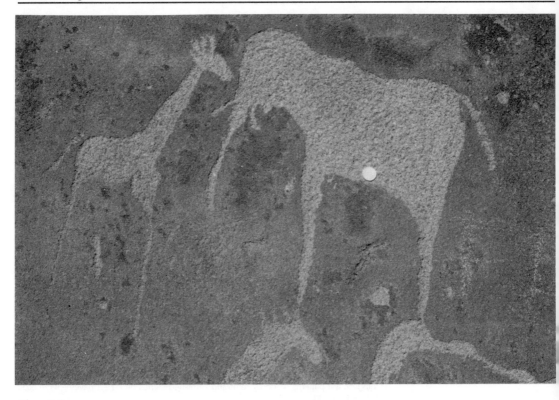

Plate 2.3
A desert varnish from Twyfelfontein, Namibia, revealed by its removal to produce rock art. Archaeological information can be used to date varnish and varnish is sometimes used to date archaeological material.

of varnish formation which include derivation from chemical weathering, derivation from aeolian sources, and derivation by bio-logical processes (Oberlander, 1994). Its significance has increased in recent years because of attempts to use it as a palaeoclimatic indicator and as a means of dating (see Dorn and Oberlander, 1982, for a review). In turn, the increasing number of dates for varnishes means that an increasing amount can be said about their rates of formation.

Early data for rates of varnish formation were often based on the estimation of ages of archaeological artefacts, a method that needs to be treated with caution, given the risk of circular argu-ment. Moreover, the attempt by Engel and Sharp (1958) to show that varnish can form in as little as 25 years has been shown to be suspect or anomalous (Hunt and Mabey, 1966; Elvidge, 1979). Elvidge (1979) has suggested that it in all probability takes about 3000 years to form a discernable varnish and some 10,000 years to form a heavy one. Indeed, as Oberlander (1994, p. 108) points out, 'Rarity of well-developed rock varnish on Holocene surfaces and artifacts suggests that in most desert environments the formation

Table 2.2 Ages of varnish deposits found by several workers with approximations as to the possible rates of varnish formation

Place	Age, years BP	Comment
Pyramids, Egypt	5000	incipient varnish, none in 2000, may take 200,000 for full patina
Arizona	60,000–200,000	for well-developed varnish
SE Utah	> 900	minimum for varnish. Old varnish = Late Wisconsin
Death Valley, California	< 10,000	on artifacts on terrace of Lake Manly
	> 2000	no varnish on artifacts less than this age
	> 50	new road formation
	< 2000	from geomorphological evidence
Barstow, S. California	25	on abandoned road (see text for comment)
Sierra Pinacate, Mexico	< 3000	unvarnished artifacts
	9000	thin varnishes on artifacts, cairns, etc.
	19,000	thick varnishes on artifacts, cairns, etc.
Grimes Pt., W. Nevada	10,000	since retreat of Lake Lahontan
Blythe, Calif./Arizona	1000	intaglios* unvarnished at this age
	500	possible date in some cases
Thorne Cave, Utah	3000	varnish on cave brow exhumed at this time

* Intaglios are patterns made by removing varnished stones on desert pavements in order
  to create a pattern.
Source: Whalley (1983), table 8.3.

of continuous black varnish coatings requires at least 10,000 years.'
Table 2.2 gives some of the ages of varnish desposits derived from
'traditional' dating methods.

With respect to rock rinds (weathering rinds), rates of formation, and the relationship between time elapsed and the rate of
weathering, can best be documented by comparing quantitative
measures of weathering with numerical estimates of time (Colman,
1981). This may involve examination of man-made structures of
known age (e.g. tombstones) or of dated geological samples (e.g.
volcanic rocks from known eruptions).

In Bohemia, Cernohouz and Solc (1966) were able to fit a loga-
rithmic curve to rind-thickness data for basalts (figure 2.2). Their
ages, which are derived from estimated ages of glacial deposits
and other sources, clearly indicate a decreasing rate of rind formation through time. They also indicate that in that environment
rinds only form slowly, a 2 mm thick rind being as much as
180,000 years old. Chinn (1981) established the rate of rind
development on New Zealand sandstone boulders associated with
radio-carbon dated deposits. His results also show that the rate of
development slows through time following an exponential curve
according to the equation

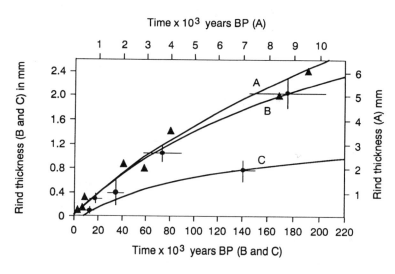

Figure 2.2
*Weathering rind thickness
versus time curves for:
A, New Zealand
B, Bohemia
C, Montana, USA
Bars represent error limits
(source: Colman, 1981,
fig. 10).*

$$t = 1030 \, d^{1.24}$$

where d is the rind thickness in mm, and t is the time in years. However, his rinds formed more quickly than those in Bohemia, with a 2 mm thick rind taking about 7000 years. In the western USA, Colman (1981) maintains that weathering rinds on basalt and andesite clasts in glacial deposits from Late Pleistocene and Holocene glacial deposits show non-linear rates of development. His data for seven areas show that the rates decline through time (figure 2.3). The rates of formation are also slow, with 2 mm thick rinds taking in excess of 250,000 years to form.

Colman (1981) attributes the decreasing rate of formation revealed in these studies to the formation of stable residues during the course of weathering. He argues that stable weathering products probably impede the flow of water to the unaltered material and slow the transport of elements away from it. Thus, he maintains, the more a material weathers, the slower it weathers, and the thicker the residual layer, the slower its thickness increases.

There is probably some climatic control on the rate of rind formation, though insufficient measurements on a sufficiently wide range of lithologies make this a difficult hypothesis to judge. Bull (1991, p. 272) has, however, suggested that rates of rind development are much more rapid in humid than in arid regions. He indicated that in the arid Mojave desert of the USA some rinds on basaltic boulders are useful for dating surfaces as old as 700,000 years, whereas in the far more humid Charwell River Valley of New Zealand, weathering rinds formed so rapidly on greywacke cobbles that their usefulness is limited to the last 20,000 years. Nonetheless, lithology is another important controlling factor, with Whitehouse, for example, finding that on the Torlesse Sandstone

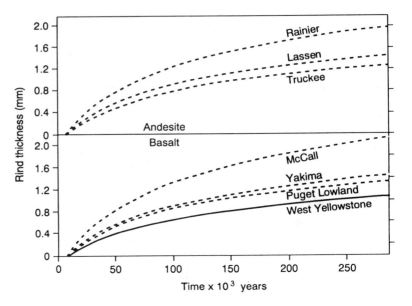

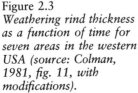

Figure 2.3
Weathering rind thickness as a function of time for seven areas in the western USA (source: Colman, 1981, fig. 11, with modifications).

of New Zealand rinds are perceptibly thicker on coarse-grained than medium-grained sandstone. Rinds were thinnest on mudstones (Whitehouse, Knuepfner, McSaveney and Chinn, 1986).

Rates of tufa development

Limestone solution in one place may lead to limestone deposition in another (plate 2.4). Indeed, in many parts of the world there may be extensive deposits of secondary freshwater carbonate (variously called tufa, travertine, sinter, etc.). Such material is geomorphologically important, creating as it does a whole suite of landforms, but it also means that in some areas published rates of current limestone denudation may be excessive in that carbonate storage may occur within the system downstream from spring water sampling points.

There is a wide range of measurement methods that have been employed to estimate tufa deposition rates, including mass balance calculations, rate law calculations, use of seed crystals or artificial substrates, use of historical records and dated layers, direct measurement using photographical and other techniques, petrographical methods, and extrapolation from annual or daily laminations (Viles and Goudie, 1990). Estimates of tufa deposition rates obtained from long-term geological evidence are likely to be prone to errors introduced by post-depositional modifications (e.g. diagenesis and erosion).

Plate 2.4
In some limestone areas, as in the Kimberley region of north-western Australia, large deposits of tufa can accumulate with some rapidity. They store calcium carbonate which would otherwise be evacuated from the system.

Table 2.3 presents some published data for rates of tufa depostion. The rates are highly variable, ranging from less than 1 mm a^{-1} to as much as 0.5 m a^{-1}.

In Europe, there is some evidence that present rates of tufa deposition are much lower than they were earlier in the Holocene. The belief is based on the fact that during the 1970s and 1980s an increasing body of isotopic dates became available for tufa from a big sample of sites. Some of these dates suggest that over large parts of Europe, from Britain to the Mediterranean basin and from Spain to Poland, rates of tufa deposition were high in the early and mid-Holocene, but declined markedly thereafter (Weisrock, 1986, pp. 165–7). The situation is demonstrated diagrammatically in figure 2.4. If this late Holocene reduction in tufa deposition is a reality then it is necessary to consider a whole range of possible mechanisms, both natural and anthropogenic (table 2.4). As yet the case for an anthropogenic role is not proven though it has many advocates (e.g. Vaudour, 1986).

Table 2.3 Selected tufa deposition rates

Location	Rate (mm a^{-1})	Method/reference
Gwynedd, Wales	0.21-0.80	Various methods (Pentecost, 1978)
Malham Tarn, North Yorkshire	0.01–1.30	Various methods (Pentecost, 1981)
Gordale Beck, North Yorkshire	1–8	Historical records (Pentecost, 1978)
Howgill Beck, North Yorkshire	0.19–0.34	Historical records (Pentecost, 1978)
Waterfall Beck, North Yorkshire	9	(Pentecost 1987)
Rivulites, various UK locations	0.16–1.59	Direct measurements (Pentecost, 1978)
Plitvice, Yugoslavia	circa 10	(Emis et al., 1987)
Tivoli, Italy	circa 8–107	Laminar counts (Chafetz and Folk, 1984)
Yellowstone Park	180	(Bargar, 1978)
Hula Valley, Israel	0.32 ± 0.10	Dated layers (Heimann and Sass, 1989)
Raclawka, Poland	10	(Pazdur et al., 1988)
Walker Lake, Nevada (Laminated rinds on shoreline)	circa 2	Laminar counts (Newton and Grossman, 1988)
Mammouth Hot Springs	20–560	(Allen, 1934)
Bee Creek, Texas	0.60	Experimental dam (Folk et al., 1985)

Source: From Viles and Goudie (1990).

The speed of formation of some other weathering phenomena

By using dated geomorphic surfaces it is possible to estimate the speed with which various other types of weathering phenomena can form. For example, Amit et al. (1993), investigated the development of shattered gravels and *reg* soils in the Negev Desert of Israel. They found that the rate of the shattering process is not linear but exponential, and becomes asymptotic with time. As figure 2.5 shows, the first stages of *reg* development and gravel shattering are very rapid, with as much as 70 per cent of the gravel shattering in the soil horizon being accomplished in no more than 14,000 years. The achievement of 80 per cent shattering takes a much longer period of time – perhaps as much as 500,000 years. Amit et al. account for this pattern of development in the following way. At first the presence of a very gravelly and permeable profile allows even quite modest amounts of salts (halite and gypsum) to cause shattering. However, with time sealing of the soil surface occurs because of the development of a pavement and the accumulation of dust and gypsum. The plugging of the soil surface serves to change the wetting depth, temperature and moisture regimes of the soil profile and at this stage of soil development the shattered gravel horizon is no longer subject to

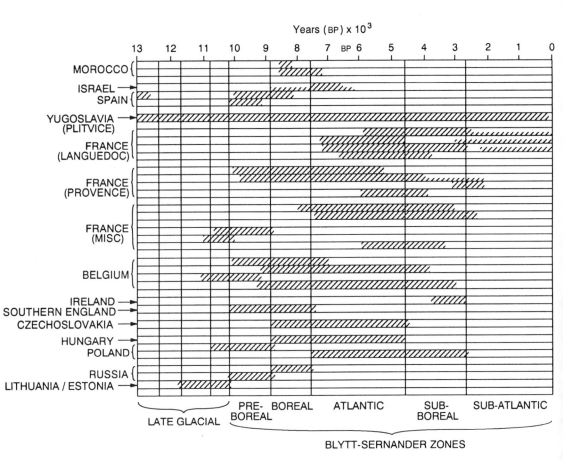

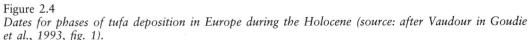

Figure 2.4
Dates for phases of tufa deposition in Europe during the Holocene (source: after Vaudour in Goudie et al., 1993, fig. 1).

the extreme conditions of wetting and drying that promote haloclasty. The salts tend to cement the gravel instead of shattering them.

The speed of tafoni and alveole development is another area where there are some data on rates of development. The normal process is to measure tafoni that have formed on man-made structures of known age (e.g. sea walls). Gill (1981), for instance, found that honeycomb weathering forms had developed on sea walls constructed of greywacke between 1943 and 1949, while Grisez (1960) found that pits developed on a schist sea wall on the French Atlantic coast at a maximum rate of about 1 mm a^{-1}. This is a rather faster rate than that measured for Jurassic oolite pillars on a seafront hotel in Weymouth, southern England (0.03 mm a^{-1}) (Viles and Goudie, 1992), but is almost exactly the same as

Table 2.4 Some possible mechanisms to account for the alleged Holocene tufa decline

Climatic/natural	Anthropogenic
Discharge reduction following rainfall decline leading to less turbulence	Discharge reduction due to overpumping, diversions, etc.
Degassing leads to less deposition	Increased flood scour and runoff variability of channels due to deforestation, urbanization, ditching, etc.
Increased rainfall causing more flood scour	Channel shifting due to deforestation of floodplains leads to tufa erosion
Decreasing temperature leads to less evaporation and more CO_2 solubility	Reduced CO_2 flux in system after deforestation
Progressive Holocene peat development and soil podzol development through time leads to more acidic surface waters	Introduction of domestic stock causes breakdown of fragile tufa structures
	Deforestation = fewer fallen trees to act as foci for tufa barrages
	Increased stream turbidity following deforestation reduces algal productivity

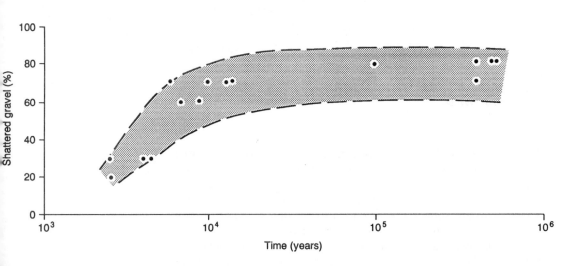

Figure 2.5
The progress of gravel shattering with time in the reg *of the Negev Desert, Israel, as a result of salt weathering (source: Amit et al., 1993).*

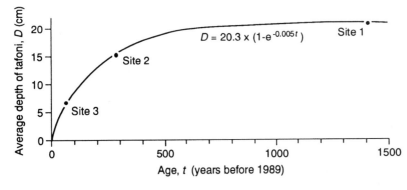

Figure 2.6
The growth of tafoni depth as a function of time (source: Matsukura and Matsuoka, 1991, fig. 5).

that determined for sea walls made of Pennant Sandstone at Weston-Super-Mare by Mottershead (1992), where 110 mm of surface reduction had taken place since construction was completed in 1886.

Data on the trend of tafoni development with time are sparse, but Matsukura and Matsuoka (1991) have, through a study of tafonis on uplifted shore platforms of known age, come up with a curve of tafoni growth as a function of time (figure 2.6). This shows that the rate of deepening of tafoni is not linear, but, like some other weathering phenomena such as rinds, an exponential function of time, with the highest rate at the initial stage. This maximum rate is about 1.67 mm a^{-1}. The reason for the exponential form of the relationship is not entirely clear, but Matsukura and Matsuoka believe that as the tafoni become deeper, so there is decreasing exposure to sun and wind, with the consequence that the desiccation required to promote salt weathering becomes rarer.

Weathering micro-forms (karren) on carbonate rocks appear capable of developing with some rapidity. For example, in south west Turkey, well developed rillenkarren have been noted on Lycian tombs and large kamenitza on fortifications at Loryma dating to about 400 BC.

The study of the weathering changes that take place in sand dunes of known age is another area where dated geomorphic surfaces provide the potential for estimating rates of chemical change. Leaching rates of shell carbonate seem to be high on coastal dunes in humid environments where organic acids are produced from decaying vegetation. On English coastal dunes, for example, Salisbury (1925) found that almost complete decalcification of the surface sands occurred within about three centuries. Likewise, in his study of the sand dunes developed from the shoreline of Lake Michigan, Olson (1958) estimated that it probably takes about 1000 years to leach carbonate fully out of the uppermost 2 m of dune sand. Mineralogical changes may also take place when carbonate rich dunes are weathered. Several studies have shown that the abundance of the relatively soluble aragonite

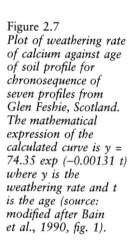

Figure 2.7
Plot of weathering rate of calcium against age of soil profile for chronosequence of seven profiles from Glen Feshie, Scotland. The mathematical expression of the calculated curve is y = 74.35 exp (−0.00131 t) where y is the weathering rate and t is the age (source: modified after Bain et al., 1990, fig. 1).

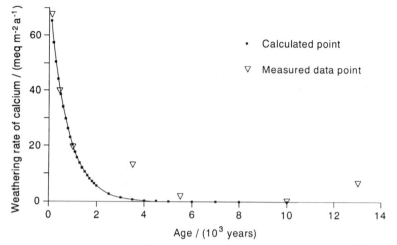

and high-Mg calcite declines with age relative to low-Mg calcite. In the case of Israeli aeolianites, Gavish and Friedman (1969) reported a total loss of high-Mg calcite within 10,000 years, and of aragonite within 50,000 years. In the case of Victoria, Australia, Reeckman and Gill (1981) found that rather longer amounts of time were involved: 100,000 years for the disappearance of high-Mg calcite and as much as 600,000 years for the disappearance of aragonite. Finally, relict and stabilized late Pleistocene desert dunes also appear to undergo substantial amounts of diagenesis in a relatively short period of time. This often involves reddening and the accumulation of fines (i.e. silt and clay). For instance, in north-western Australia, late Pleistocene dunes, formed under more arid conditions when the Great Sandy Desert extended further to the north than it does today, have been reddened to depths of at least 6 m and have silt and clay contents that are typically around 20 to 25 per cent (compared with less that one per cent for currently active dunes further to the south). SEM, XRD and petrographic investigations indicate that a large portion of the fines content of the relict dunes is derived from weathering of sand-sized feldspars and possibly other alumino-silicate minerals, and that much of the dune reddening took place *in situ* due to crystallization of haematite (Goudie et al., 1993).

Another example where a suite of geomorphological features can be used to assess trends in weathering rates through time is the study of soil development on sequences of river terraces of different ages. In the case of Glen Feshie in Scotland (figure 2.7), weathering rates were determined over a period of 13,000 years, by comparing the loss of cations in soil profiles of different ages relative to a stable mineral (Zr). In the case of calcium, the rate of loss appears to decrease exponentially with time, a pattern that applies to many weathering processes and phenomena. This tendency for some weathering phenomena to decelerate through time was deduced qualitatively by Penck (1953, pp. 50–1):

It is not known how long a time is required for any particular rock to become loosened, and then to develop a rubbly zone and finally an horizon of purely end-products. Therefore it is not known how long a period of reduction is needed for soil horizons to develop from one another in the direction of an increasing degree of reduction. All that is certain is that it takes considerably more time for the unmixed end-product to develop from the zone of coarse rubble than for the latter to arise from the zone of loosening. The following statement makes this clear: the task assigned to weathering may be expressed by the frequency of division necessary for cutting the material up into smaller and smaller fragments (i.e. by the number of the divisional planes). Thus it is obvious, that the mere loosening of the rock is a far smaller task than the multiple subdivision in the rubbly horizon, and this again is one immeasurably less than the decomposition of the material into colloids or very fine dust, which means an almost infinite multiplication of the process of division. The number of divisional surfaces grows in geometrical progression. If such an ever increasing task were to be performed in consecutive equal periods of time, the intensity of the reducing processes would have to be raised to the power of the number of such periods. But, physical and chemical agents taken together have, on the whole, a constant intensity at a given place. Therefore it follows that, compared with the latter phases, the initial stages of rock reduction are actually passed through far more quickly: that it takes a shorter period of reduction to loosen rock than to change the zone of loosening into one of rubble: and that this latter again arises much more rapidly than the end-product develops from it – *always assuming similar exposure. The degree of reduction is not proportional to the duration of the reduction; but, the exposure being the same, decelerates.*

Rates of weathering determined from laboratory studies

Weathering has always been one of those areas in geomorphology where experimental studies in the laboratory have been seen as apposite. Indeed, laboratory-based studies of chemical and other processes of breakdown have a long history, with many notable studies being conducted in the early decades of the nineteenth century. There are, inevitably, certain limitations to what can be achieved in the laboratory and to the degree of confidence one can have in extrapolating laboratory results to the real world of the field situation. For example, most experimental studies of chemical change use simple mineral and chemical environments, and do not adequately replicate the sorts of biological and hydrological conditions encountered in real weathering environments.

Nonetheless, experiments in the laboratory have served to demonstrate the rapidity with which certain changes can occur given optimal conditions: laboratory simulations, for example, did much to gain acceptance of the power of salt weathering as a process (see, for example, Sperling and Cooke, 1985), the role of concentrated salts in promoting silica translocation (see, for example,

Magee, Bull and Goudie, 1989), and the role of organic acids in promoting rapid changes in dune sand profiles (see, for example, Williams and Yaalon, 1977).

Laboratory studies have also served to demonstrate the relative susceptibilities and resistances of different minerals under different conditions. A useful instance of this is Lasaga's (1984) study of the comparative rate of dissolution of silica from different minerals at a standard temperature (25°C) and acidity (pH 5). His estimates of the average time required for the complete dissolution of 1 mm crystals were as follows:

Mineral	Years
Quartz	34,000,000
Muscovite	2,700,000
Forsterite	600,000
K-Feldspar	520,000
Albite	80,000
Enstatite	8,800
Diopside	6,800
Nepheline	211
Anorthite	112

Biological weathering rates

Despite widespread acknowledgement of the importance of many different types of organisms (e.g. bacteria, algae, lichens, fungi, snails and plant roots) to rock weathering there have in fact been only a few meaningful measurements (table 2.5) so far. There are several reasons for this paucity of information. Firstly, it is often difficult to identify specific biological weathering processes and then to disentangle them from other inorganic mechanisms. Micro-organisms which grow in the chasmoendolithic niche (i.e. under the rock surface, exploiting pre-existing pore spaces), for example, may provoke rock breakdown, but it is difficult to isolate their role from other processes such as exfoliation and chemical reactions. Secondly, many organisms are capable of carrying out a range of weathering processes over their life span and may also act to protect rock surfaces as well. Thus many saxicolous lichen species exert chemical and mechanical stresses on the underlying rock, but also act as a protective 'biofilm' over the surface. Thirdly, rates of biological weathering vary enormously over time as community development (succession) takes place. Any analysis of the geomorphological importance of biological weathering must take into consideration successional changes. Most short-term weathering studies will only observe a fraction of any rock surface succession. Finally, there are several practical difficulties involved in monitoring biological weathering. Laboratory experimentation is virtually impossible for some organisms (e.g. lichens and trees).

Table 2.5 Biological weathering rates

Place	Organisms	Rates	Notes/source
Limestone in the Negev desert (90 mm of mean annual rainfall)	Snails and endolithic lichens	0.7–1.1 tonnes ha^{-1} a^{-1} (0.63–0.99 m^3 ha^{-1} a^{-1})	Schachak et al. (1987)
Limestone in W. Malaysia (more than 2400 mm of mean annual rainfall)	Tropical forest	600–729 kg ha^{-1} a^{-1}	Ca and Mg uptake by plants, Crowther (1987)
Limestone walls in Jerusalem (mean annual rainfall 611 mm)	Cyanobacteria	0.005 mm a^{-1}	Danin (1983)
Granitic nunataks, Juneau Icefield, Alaska	Chasmolithic green algae	562 g m^{-2} loose material produced – 0.005–0.02 mm a^{-1}	Hall and Otte (1990)
Sandstone wall in Orange Free State, South Africa	Endolithic lichen	0.096 mm a^{-1}	Wessels and Schoeman (1988)

Even where field or laboratory experiments can be performed (for example, using endolithic algae that perforate rocks) it can be difficult to translate the results into a weathering rate (e.g. number of boreholes per cm^3). Special field measurement techniques often need developing for biological weathering processes – such as measuring the calcium content of snail faeces – as standard methods do not pick up many esoteric biological effects.

Looking at table 2.5 it is apparent that most measurements of biological weathering come from 'extreme' environments, specifically deserts, periglacial and tropical rainforest areas.

Accelerated weathering in cities

The high levels of atmospheric pollutants in cities, especially sulphates and nitrates, mean that susceptible building materials may be subjected to acid attack, suphation and other weathering processes. Quantitative data on the rates at which such urban weathering occurs have been obtained from sources which include tombstones and lead plugs in walls (see, for example, Trudgill et al. 1991; Cooke et al. 1993), weight loss from specially prepared stone samples, chemical analysis of runoff from exposed stone surfaces, and micro-erosion meter studies.

Given that pollution rates have changed through time in response to industrialization, industrial decline, the growth in road traffic, and changes in fuel type and use, it might also be anticipated that urban weathering rates might also have changed through time. It is, however, possible that weathering rates at any one time are strongly influenced by antecedent weathering conditions, creating a lag. This has been termed 'the memory effect', and may occur when, for example, sulphates, resulting from domestic coal burning, continue to cause stone decay even when the use of coal fires is abandoned. Moreover, older headstones may experience an inherently accelerated decay rate as the weathering process increases the stones' porosity, permeability and surface area available for attack.

Tombstones have proved to be especially useful to determine changing rates: they are ubiquitous, they are well dated, they create similar micro-weathering environments, they have a relatively wide range of ages, and may be made of similar (and thus comparable) rock types. Quantitative data can be obtained from them by an examination of their lead lettering. As Attewell and Taylor (1988, p. 748) have pointed out:

The lettering on the gravestones is formed by the stonemason carving out the shape of the letter in the marble and then filling the groove with molten lead. The lead is then finished flush with the surface of the marble, leaving the dark grey lead visibly prominent against the marble background to record the important date when the headstone was placed in the graveyard.

Since the lead lettering is relatively unaffected by pollutants in the atmosphere, the marble decays with time to leave the lettering protruding from the headstone surface. The rate of weathering can be derived by measuring the distance between the lettering and the marble for a range of gravestones of various ages.

In addition quantitative rates can be obtained by analysis of inscription legibility (Meierding, 1993).

Some rates for tombstone weathering from polluted urban environments are shown in table 2.6. The rates range from 2 to 139 microns per year for limestones and marbles. Studies tend to show that the depth of material that is lost increases with age. This is to be expected given that the longer the gravestone is exposed the more material is likely to be liberated by weathering. However, in the case of examples from Britain different cities display different rates (Cooke et al., 1993) (figure 2.8a, b, c) and some show an acceleration of weathering rate with age, though not all. In the case of Haifa in Israel, Klein (1984) discovered an increasing rate of weathering with time (figure 2.8d) and found that an exponential curve gave a better fit to his data than a linear one (an assertion that was later contested by Neil, 1989). However, Klein cautioned against assuming that this upward trend was caused by rapid urbanization, and suggested that the changing rate could be the result of tombstones loosing their polish through time. He explained (p. 108): 'At first, the stone is nicely polished and the rate of weathering is low; after a few decades, the stone becomes

Table 2.6 Rates of weathering determined from the degree by which lead lettering stands out from tombstones

		mm 1000 a^{-1} units	
Philadelphia			
City Centre[2]	Marble	35	Feddema and
20 km from centre[3]		5	Meierding (1987)
Eastern Australia			
Wollongong[1]	Marble	2.5	Dragovich (1986)
Sydney[2]		1.7	
Israel			
Haifa[2]	Limestone	5	Klein (1984)
England			
Co. Durham	Marble		Attewell and Taylor (1988)
Meadowfield[2]		10	
Quebec Village[3]		2	
Jarrow[2]		9.5	

Note:
[1] = Industrial.
[2] = Urban.
[3] = Rural.

unpolished and the rate of weathering increases.' Nonetheless, Feddema and Meierding (1987) using old photos of gravestones in Philadelphia, USA, and comparing them with modern repeat photographs, found a marked acceleration in exfoliation between 1930 and 1960 which was concurrent with greatly increased atmospheric SO_2 levels in the city during that period. Support for their contention also came from a comparison of tombstone weathering rates from urban and rural sites in Philadelphia and its environs.

The centre of Philadelphia showed mean recession rates for marble tombstone surfaces of 3.4 mm 100 a^{-1}, whereas, less polluted rural rates were over six times lower, tending to be below 0.5 mm 100 a^{-1}. Meierding (1993a), using tombstones made of Vermont marble, and employing his 12-point scale of inscription legibility, found that in some areas of North America (e.g. small towns in the Great Plains, parts of the south, and the Maritime Provinces of Canada) rates were less than 0.1 mm 100 a^{-1} whereas in some coal-burning areas they were very considerably higher (e.g. Western Pennsylvania, 1.31 mm 100 a^{-1}, and north Missouri, 1.64 mm 100 a^{-1}). Similarly the studies of Cooke et al. (1993) in Swansea (Wales) and of Dragovich (1986) in Wollongong (Australia) have shown that the presence of smelters leads to especially rapid rates of stone decay (figure 2.8e and f). In the latter case Dragovich found that the mean rate of surface reduction in the industrial environment (with smelter and steelworks) was 0.25 mm 100 a^{-1} compared with 0.17 mm 100 a^{-1} in a more residential area. Furthermore the rate of weathering accelerated over time in

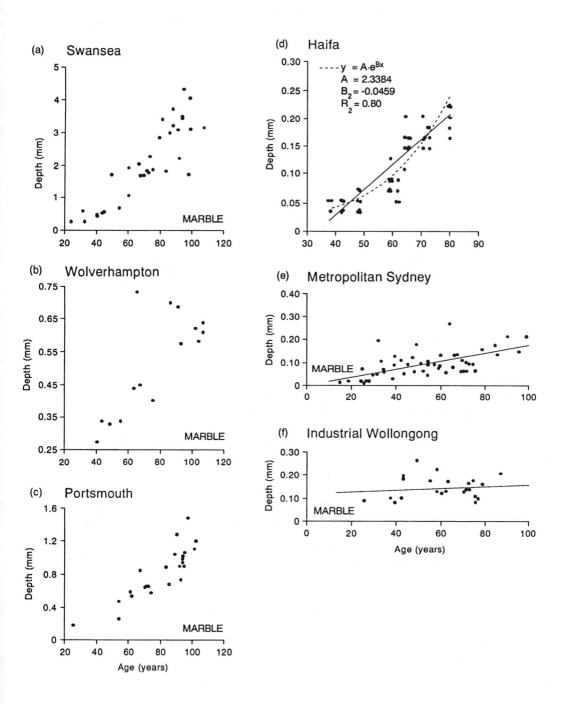

the industrial area but not in the more residential area (figure 2.9). In the case of Durham in north-east England, Attewell and Taylor (1988) also found differences between industrial and rural areas. A rural cemetery (Quebec) had a weathering rate of 0.2 mm 100 a^{-1}, whereas Consett (a steel-making centre) had a rate of 1.15 mm 100 a^{-1}, an almost six-fold increase on the rural rate.

One of the most comprehensive attempts to ascertain regional patterns of weathering related to air pollution is Meierding's (1993) study of marble tombstone weathering in North America. Using data from 320 cemeteries and over 8000 Vermont marble tombstones he found that degradation rates were negligible in dry areas (e.g. the western USA and the High Plains); rainy, less polluted areas (e.g. Hawaii and the south-eastern United States); cold regions (e.g. Canada and high altitude Rocky Mountains); and near coal-fired power plants with tall stacks. Weathering rates have been rapid where charcoal or high-sulphur coals have been used as fuels (e.g. industrial cities of the eastern USA, small towns in the mid-west and near oil smelters).

Estimation of chemical denudation rates from river water analyses

Having in the first part of this chapter discussed the rates at which various weathering phenomena develop and weathering processes operate, the second part of this chapter deals with the major issue as to what one can learn about rates of chemical denudation through a study of river water flow and chemistry.

In estimating rates of solutional denudation it is extremely important to make allowance for the non-denudational component of the dissolved load, for substantial amounts of material may be introduced into a catchment in precipitation. It is significant that if one examines the average composition of world river water two major components are Na^+ and Cl^-, the two most abundant dissolved constituents in sea water. Also of great significance is the presence of HCO_3^-, which arises from the incorporation of atmospheric CO_2 during weathering reactions. Indeed, it has been

---

Figure 2.8
Rates of weathering of gravestones in urban environments:
(a) Swansea, south Wales. Note the high rate which may be associated with smelter effects.
(b) Wolverhampton, an industrial centre in the English Midlands.
(c) Portsmouth, a coastal city in southern England.
(d) Haifa, a coastal city in Israel.
(e) Metropolitan Sydney, Australia.
(f) Wollongong, an industrial city in S.E. Australia.
(sources: Cooke et al., in press; Klein, 1984, fig. 1 with modifications; and Dragovich, 1986, fig. 3 with modifications).

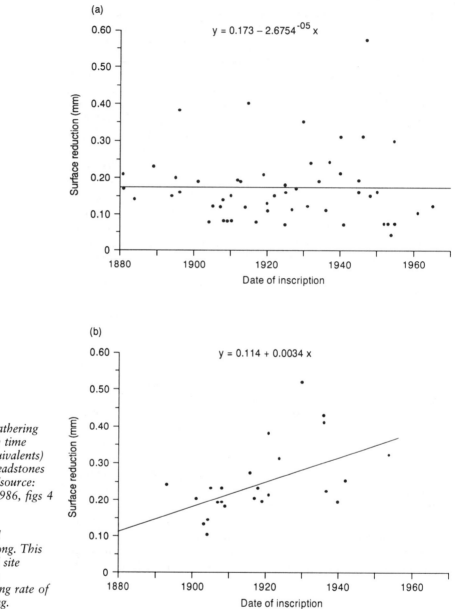

Figure 2.9
*Chemical weathering
rates through time
(100-year equivalents)
for marble headstones
in Australia (source:
Dragovich, 1986, figs 4
and 5):*
(a) Sydney.
*(b) Industrial
    Wollongong. This
    industrial site
    shows an
    accelerating rate of
    weathering.*

calculated (see table 2.7 and Summerfield, 1991a, p. 382) that the
non-denudational components of river runoff amount to about 40
per cent of the solute load of rivers on a global basis.

The significance of the non-denudational component varies according to catchment conditions. When rates of rock weathering
are high because of the availability of large amounts of moisture and
highly reactive rock types (e.g. limestone, evaporites, etc.) the non-
denudational component will be relatively modest in importance.

Table 2.7 Average composition of world river water and estimates of denudational and non-denudational contribution for different constituents

	Ca^{2+}	Mg^{2+}	Na^+	K^+	Cl^-	SO_4^{2-}	HCO_3^-	SiO^2	Total
Average composition of world river water conc. (mg l^{-1})	13.5	3.6	7.4	1.35	9.6	8.7	52.0	10.4	106.6
Provenance of major solute components (%) Non-denudational: Precipitation (Oceanic salts)	2.5	1.5	5.3	14	72	19	–	–	12
Atmospheric CO_2	–	–	–	–	–	–	57	–	28
Denudational Chemical weathering	97.5	85	47	86	28	81	43	100	60

Source: Based largely on data in M. Meybeck (1983) in: *Dissolved Loads of Rivers and Surface Water Quantity/Quality Relationships*, International Association of Hydrological Sciences Publication, 141, 173–92, in Summerfield (1991) table 15.4.

By contrast in areas that are already heavily leached, covered with chemically inert weathering mantles, and underlain by non-reactive lithologies (e.g. quartzites) the non-denudational component may take on very considerable significance. For some ecosystems and for some elements, the stream output may be less than the atmospheric inputs (Meybeck, 1983, p. 176).

Also highly influential in terms of atmospheric inputs is proximity to the oceans, these being the major source of atmospheric material. Meybeck (1983, p. 178) suggests that when the distance to the coast exceeds 100 km, the oceanic influence is gradually superseded by continental inputs, resulting from such sources as soil dust and forest fires.

There is, however, in general a decrease in the dissolved content of rain as one moves inland. If the precipitation amount is constant, the Cl^- content usually decreases exponentially in two or three steps, which correspond to the fallout of firstly the larger sea-spray particles and then of smaller salt nuclei. In the Mediterranean regions of France the Cl^- input to land exhibits a 10-fold decrease in the first 100 km inland from the sea (Meybeck, 1983, p. 180).

Inputs of material in precipitation have a geomorphological significance beyond their role in providing the non-denudational component of river water chemistry. They may also have a significance in terms of rock weathering (e.g. through the provision of soluble salts for salt weathering) and in terms of the formation of miscellaneous types of chemical sediment (e.g. calcretes, salt crusts, etc.). In systems that are strongly influenced by anthropogenic pollution, rates of rock weathering may be accelerated by 'acid precipitation'.

Making allowance for the role of atmospheric inputs, while desirable and necessary (Meade, 1969; Goudie, 1970; Janda, 1971), is a difficult procedure. Because of the tendency for recycling to occur within a basin it is not easy to differentiate it from material that comes in from outside the catchment. However, by analyzing precipitation inputs and by using available knowledge of bedrock chemistry, some realistic adjustments may be made. Walling and Webb (1978), for instance, proposed that in their catchments in Devon, south-west England, there were several ionic constituents that were likely to have been derived from non-denudational sources. These were subtracted from the total output figure to gain a measure of net denudation. They also calculated the theoretical atmospheric contribution to solute loads by making allowance for concentrations of the atmospheric input by evapotranspiration:

$$C_s = C_pP/(P-ET)$$

where C_s is the theoretical solute concentration in streamflow attributable to atmospheric sources (in mg l^{-1}), C_p is the average solute concentration of bulk precipitation in the same units, P is the annual precipitation quantity (mm) and ET is the annual rate of evapotranspiration (also in mm). The associated non-denudational load component (L_n) in t km^{-2} can be calculated as

$$L_n = RC_s/1000$$

where R is the annual runoff in mm (i.e. P–ET).

The results of these two methods were broadly comparable in that they both illustrated the substantial non-denudational component of the stream dissolved loads (table 2.8). Such non-denudational components are also becoming increasingly important as a consequence of human inputs into drainage systems. These sources include the addition of nitrate fertilizers, acid precipitation and municipal and industrial wastes. The calculation of these types of inputs is also fraught with problems (Kostrzewski and Zwolinski, 1985).

A problem that needs to be considered in estimating rates of chemical denudation from river water analyses is the sometimes arbitrary distinction that is normally drawn between the dissolved and suspended loads. For example, colloidal material may be viewed as extremely fine sediment (i.e. diameter less than 45 µm), but in most laboratory procedures will not be retained on the filter medium and so will be included in the solutes represented by the filtrate (Walling, 1984). Furthermore, during transport there may be interchanges between the dissolved and solid phases as a result of various chemical mechanisms (e.g. solution, precipitation, adsorption and desorption).

Table 2.8 Estimates of chemical denudation rates for gauged catchments in Devon

Site	Gross denudation	Chemical denudation[+]	(%)[*]	Chemical denudation[++]	(%)[*]
1	35	23	65.7	26	74.3
2	11	0	0	6	54.5
3	24	14	58.7	16	66.7
4	28	17	60.7	20	71.4
5	20	7	35.0	13	65.0
6	35	26	74.3	27	77.1
7	26	15	57.7	18	69.2
8	23	15	65.2	17	73.9
9	39	32	82.1	29	74.4
10	32	22	68.8	22	68.8
11	24	18	75.0	19	79.2
12	23	16	69.6	17	73.9
Mean	–	–	59.4	–	70.7

[+] Chemical denudation estimated with non-denudational component derived from precipitation chemistry data (m^3 km^{-2} a^{-1}).
[++] Chemical denudation estimated with non-denudational component derived from chemical composition of total stream load (m^3 km^{-2} a^{-1}).
[*] Percentage of gross denudation.

Dissolved loads of rivers

Bearing all the above caveats in mind, analysis of river discharge and the concentration of dissolved material in such flow permits one both to establish rates of chemical denudation and to establish the relative efficiency of mechanical and chemical erosion. Studies have ranged in scale from global assessments of loads entering the oceans to detailed investigations of solute movements in small catchments.

At the global scale, a detailed analysis has been undertaken by Walling and Webb (1983), who built up a data base for 490 rivers 'in which pollution was of limited significance' (p. 4). The authors admit that the database is not entirely representative of the global situation, as certain parts of the world (notably regions of South America and Africa) are short on data. The frequency distribution of the 490 values of mean annual dissolved load is illustrated in figure 2.10. The mean load was found to be 38.8 t km^{-2} a^{-1}, and typically values lie in the range 5–100 t km^{-2} a^{-1}. There were some values below 1.0 t km^{-2} a^{-1} (for small catchments in Alberta), while the highest value was for the River Iller in Germany (311 t km^{-2} a^{-1}).

The data set also demonstrated a general influence of climate in that there was a positive relationship, albeit rather weak, between annual dissolved load (D) and mean annual runoff (Q), expressed by this least squares regression:

$$D = 3.3Q^{0.385} \quad (r = 0.49)$$

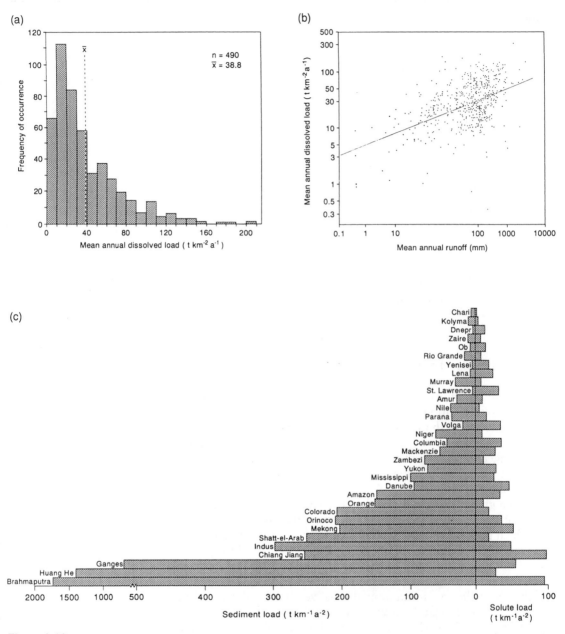

Figure 2.10
River dissolved loads:
(a) The frequency distribution of the values of mean annual dissolved load in the Walling and Webb data base (source: Walling and Webb, 1983, fig. 1).
(b) The relationship between mean annual dissolved load and mean annual runoff from the same data base (source: Walling and Webb, 1983, fig. 2).
(c) Solute loads and sediment loads for the world's largest drainage basins (source: Summerfield, 1991a, fig. 15.11).

The explanation for the positive trend may be that increasing moisture availability increases the rate of chemical weathering and solute evacuation. The explanation for the poor level of explanation of the regression is that many other factors control solute delivery, including temperature, seasonality, rock type, and vegetation. Of these, rock type may well be the most important for the dissolved loads of rivers draining igneous rocks are, not surprisingly, lower than those found in areas of sedimentary rocks.

A third major finding arising from analysis of the database was that the dissolved load transport to the oceans on a global scale is considerably less than particulate transport. The ratio is about 3.6:1. Values of the ratio range from in excess of 100 to less than 0.5, and the particulate component exceeds the dissolved component in more than 60 per cent of cases. Given that the chemical load may include a large precipitation input component, and given that large amounts of particulate sediment may not be delivered right through the catchment and into the oceans, Walling and Webb (1983, p. 16) believe that mechanical erosion may be of considerably more importance in landscape development than is indicated by a simple comparison of particulate and dissolved loads. Indeed, they believe that the ratio of 3.6 may need to be increased 'by an order of magnitude in order to produce a meaningful estimate of the relative importance of mechanical and chemical erosion' (p. 17).

Schmidt (1985) undertook a comparison of mechanical and chemical denudation in the Upper Colorado river system of the south-west USA, based on 42 catchments (figure 2.11). He found some clear relationships between the two different types of denudation and the amount of runoff in a climatic zone where mean annual precipitation ranged from 150–1500 mm. Chemical denudation increases with runoff as does its percentage of total denudation. Mechanical denudation rate increases as runoff increases. The importance of this is that in areas with low runoff soluble rocks are relatively resistant, helping to explain the tendency for limestone regions to form highlands in the semiarid zone, whereas in some limestone regions in the humid zone topographic depressions are developed.

High rates of chemical denudation are invariably observed in high mountainous regions (Summerfield, 1991a, p. 385), though the accompanying high rates of mechanical denudation prevent the accumulation of deep weathering profiles. By contrast minimal rates of chemical denudation are recorded in dry regions where runoff is low (table 2.9). Rivers that drain largely semi-arid regions, such as the Colorado, Orange and Tigris-Euphrates (Shatt-el-Arab), have low rates of solute transport in relation to total transport. This contrasts markedly with basins from humid and subarctic regions such as the Lena, Yenisei, Ob and Dnepr which have high rates of solute transport in relation to total transport. In the case of the St Lawrence River, the relatively high proportion

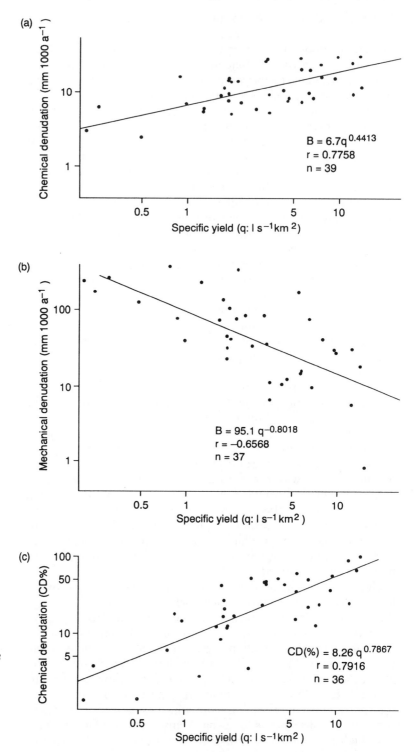

Figure 2.11
Relationships between denudation rate and runoff (specific yield) (source: Schmidt, 1985, figs 3, 4 and 5).
(a) Relationship between specific yield and chemical denudation (mm 1000 a^{-1}).
(b) Relationship between specific yield and mechanical denudation (mm 1000 a^{-1}).
(c) Relationship between specific yield and chemical denudation as a percentage of total denudation.

Table 2.9 Estimated denudation rates for the world's thirty-five largest drainage basins based on solid and solute transport rates

	Drainage area (10^6 km^2)	Total denudation (mm 1000 a^{-1})	Chemical* denudation (mm 1000 a^{-1})	Chemical denudation as % of total
Amazon	6.15	70	13	18
Zaire (Congo)	3.82	7	3	42
Mississippi	3.27	44	9	20
Nile	2.96	15	2	10
Paraná (La Plata)	2.83	19	5	28
Yenisei	2.58	9	7	80
Ob	2.50	7	5	70
Lena	2.43	11	9	81
Chiang Jiang (Yangtze)	1.94	133	37	28
Amur	1.85	13	3	22
Mackenzie	1.81	30	10	33
Volga	1.35	20	13	64
Niger	1.21	24	11	47
Zambezi	1.20	31	3	11
Nelson	1.15	–	–	–
Murray	1.06	13	2	18
St. Lawrence	1.03	13	12	89
Orange	1.02	58	3	5
Orinoco	0.99	91	13	14
Ganges	0.98	271	22	8
Indus	0.97	124	16	13
Tocantins	0.90	–	–	–
Chari	0.88	3	1	29
Yukon	0.84	37	10	28
Danube	0.81	47	16	35
Mekong	0.79	95	20	21
Huang He (Yellow)	0.77	529	11	2
Shatt-el-Arab	0.75	104	11	11
Rio Grande	0.67	9	3	38
Columbia	0.67	29	13	46
Kolyma	0.64	5	2	31
Colorado	0.64	84	6	7
São Francisco	0.60	–	–	–
Brahmaputra	0.58	677	34	5
Dnepr	0.50	6	5	88

* Allowance made for non-denudational component of solute loads.
Source: modified after Summerfield (1991a) table 15.6.

carried by solute transport may be attributed to the large proportion of particulate sediment trapped in the Great Lakes.

The observation that high rates of chemical denudation occur in mountainous regions is a significant one, and Summerfield (1991a, p. 387) believes that 'rates of chemical denudation appear to be much more strongly related to relief than to climate'. He illustrates this by reference to the Amazon basin, where around 85 per

Table 2.10 The geographical origins of dissolved loads to the oceans and variations of chemical erosion according to morphoclimatic regions (after Meybeck, 1979)

Climatic region	Area of exoreic runoff[1]	Exoreic runoff[2]	Transport of silica		Transport of ions		Chemical erosion (t km^{-2} a^{-1})
			Dissolved load to oceans (10^6 t a^{-1})	%	Dissolved load to oceans (10^6 t a^{-1})	%	
Tundra and taiga	20.0	10.7	15.0	3.9	466	13.1	14
Humid taiga	3.15	3.4	5.0	1.3	74	2.1	15.5
Very humid taiga	0.2	0.6	1.1	0.2	9	0.25	32
Pluvial temperate	4.5	15.3	45	11.8	540	15.4	80
Humid temperate	7.45	7.75	17.5	4.6	407	12.0	35
Temperate	6.7	3.35	9.4	2.5	301	8.8	28
Semi-arid temperate	3.35	1.05	2.7	1.0	130	3.7	24
Seasonal tropical	13.25	5.85	31.1	8.2	119	3.4	6.4
Humid tropical	9.2	8.85	38.2	10.2	239	8.0	15.5
Very humid tropical (plains)	6.9	18.45	78.6	20.8	165	4.8	22
Very humid tropical (mountains)	7.95	24.05	130	34.4	908	25.6	67
Total of tropical zone	37.3	57.2	278	73.6	1431	41.8	
Arid	17.2	0.65	3.9	1.0	3457	2.8	3
Pluvial regions of strong relief	12.65	40	176	47	1457	42	≈ 74

Note:
[1] = % of total land surface area.
[2] = % of total exoreic runoff quantity.

cent of the total solute load comes from the Andean headwaters (see also, Gibbs, 1967).

Meybeck (1979) (see table 2.10) attempted to classify rates of chemical denudation according to some major climatic zonations. As can be seen, the highest rates occurred in 'pluvial temperate' zones, many of which have very high rainfall amounts associated with high relief conditions. Only slightly less important was the very humid tropical mountainous zone. Not surprisingly, the lowest rates occurred in dry environments (tundra and taiga, the seasonal tropics, and arid lands).

At a more local scale, Dunne (1978) examined the controls of rates of chemical denudation on silicate rocks (ranging from rhyolites to basalts) in Kenya, and discovered that there was a very clear relationship between mean annual runoff and the removal of dissolved solids in streams.

Great interest has been attached by geomorphologists to the rate at which silica is evacuated in the dissolved form from humid tropical catchments. The presence of deeply weathered profiles, and in particular the presence of extensive laterite (ferricrete) formations, has traditionally been thought to indicate very high rates of silica removal in tropical rivers. However, the evidence is not especially conclusive.

This was appreciated by Merrill (1904, p. 278), who even that long ago was urging caution with respect to notions of the very rapid rate of chemical denudation in the tropics: 'For many years an impression has prevailed to the effect that rocks decomposed more rapidly in warm and moist than in cold climates. While, owing to abundance of vegetation and other supposed favourable conditions, a more rapid decomposition might be expected, such has not yet been proven to actually take place, and indeed many facts tend to prove the impression quite erroneous.'

Douglas (1969) approached this problem by looking at water analyses that he had obtained from his own field research in Queensland and Malaysia, and also by a survey of some pre-existing analyses from other tropical and extra-tropical rivers. Table 2.11 shows some of the data he employed. What is evident from these data, which have been modified from the original source to show the percentage that silica makes of the total dissolved solids concentrations, is that silica makes up a high percentage of the dissolved concentrations in these tropical rivers (averaging over 27 per cent for the tropical rivers, and 24 per cent for the small tropical catchments in Malaysia and Queensland). By contrast the percentage for the extra-tropical rivers only averages about 4.6 per cent. On the other hand, the actual concentrations of silica in the two climatic zones do not appear to be greatly different. When Douglas analysed silica loads of tropical and extra-tropical rivers (as opposed to silica concentrations) runoff was found to be the main control (table 2.12). He asserted (p. 12) that 'the temperature factors affecting weathering rates and thus silica concentrations in drainage waters are less important in total geomorphological work done in removing silica than contrasts in precipitation and other factors affecting run-off'. He also suggested that the presence of thick weathering residues and the like might possibly be better explained not by high rates of chemical denudation but by periods of long stability and slow landform change under a well-established forest cover. Subsequent work has suggested that large amounts of dissolved material are tied up in the vegetation mantle rather than bring released into drainage waters.

This is a point that has been taken up by Bruijnzeel (1983, p. 230): 'The chemical flux from a drainage basin will only represent the ongoing chemical denudation rate if the vegetation has reached a steady state with respect to uptake and a return of nutrients. As long as there is a net increase in biomass, a proportion of the chemical elements released by weathering will be incorporated in the vegetation and the chemical flux will underestimate the true denudation rate.' However, Bruijnzeel also undertook a comparative analysis of silica loadings in tropical and temperate streams (table 2.13), and in contrast to Douglas (1969) found the concentrations of silica in the former were significantly greater than in

Table 2.11 Silica and total dissolved solids concentrations for tropical and extra-tropical rivers

	Silica (mg l^{-1})	Total dissolved solids (mg l^{-1})	Silica as % of total	
Tropical rivers				
Mekong	15	198	7.58	
Niger	10	49.1	20.37	
Senegal	n.d.	63.4	–	
Orinoco	8	53.5	14.95	x = 27.10
Essequibo	15.8	34.1	46.33	
Demerara	40.8	81.5	50.06	
Amazon at Obidos	10.6	43.1	23.34	
Amazon at mouth	n.d.	36	–	
Extra-tropical rivers				
St Lawrence	5.7	166	3.43	
Hudson	4.9	173	2.83	
Mississippi	5.9	221	2.67	
Rio Grande	30	881	3.41	
Columbia	13	191	6.81	
Yukon	13	268	4.85	x = 4.57
Mackenzie	3.4	219	1.55	
Rhine	n.d.	598	–	
Elbe	8.6	892	0.96	
Volga	12.5	458	2.73	
Plate	19.4	103	18.83	
Thames at Walton	8.1	368	2.20	
Small tropical rivers in Malaysia and Australia				
Tebrau (W. Malaysia)	8.2	35	23.43	
Scudai (W. Malaysia)	9.8	45	21.78	
Gombak (W. Malaysia)	15	75	20.00	
Barron (Queensland)	12	65	18.96	
Behana (Queensland)	12	34	35.29	x = 24.31
Freshwater (Queensland)	14	52	26.92	
Mary (Queensland)	7	43	16.28	
Millstream (Queensland)	10	45	22.22	
Wild (Queensland)	14	66	21.21	
Nitchaga (Queensland)	18	48	37.50	

Source: modified after Douglas (1969) table II.

the latter. Average values were 18.1 ± 4.7 and 5.0 ± 2.3 mg l^{-1} respectively. Plainly there is a need for greater data in this area and for a consideration of other non-climatic controls (e.g. lithology) before any firm statement can be made about the differences in silica concentrations and loadings between different climatic environments.

Meybeck (1979) also showed that rates of silica removal were especially high in tropical regions. Indeed, he maintained that on a global basis something like three-quarters of the silica transported to the oceans in river water came from the tropical zone (figure 2.12).

It has often been surmised that rates of chemical denudation

Table 2.12 Average silica concentrations, runoff and mean annual silica loads of tropical and extra-tropical rivers

Catchment	Area (km²)	Average silica concentration (ppm)	Mean annual runoff (mm)	Mean silica annual transport (mm 1000 a⁻¹)
A. Tropical Rivers				
Behana Creek (Qld)	83	12	2325	11.01
Sg. Gombak (Malaya)	140	15	856	5.92
Freshwater Creek (Qld)	44	14	986	4.73
Nitchaga Creek (Qld)	75	18	818	4.57
Me (Ivory Coast)	4000	15	393	2.4
Barron (Qld)	225	12	657	2.27
Mary (Qld)	91	7	1685	2.05
Millstream (Qld)	92	10	621	1.80
Wild (Qld)	585	14	421	1.51
Agneby (Ivory Coast)	8700	14	180	1.0
x		13.1		3.73
B. Extra-tropical rivers				
Cowlitz (Wash.)	5796	14	1618	8.57
Chehalis (Wash.)	3352	16	1172	6.84
Anchor (Alaska)	585	28	437	4.19
Sawmill Creek (Alaska)	100	2	4340	3.5
Kenai (Alaska)	164	5	1800	3.3
Bramina Creek (NSW)	68	10	738	3.21
Trail (Alaska)	505	4	1568	2.32
Loire (France)	30,000	31	300	1.8
Brindabella Creek (NSW)	26	6	391	1.64
Johnston's Creek (NSW)	3	3	1125	1.18
Second Creek (Minn.)	68	16	225	1.02
Mittagong (NSW)	1	4	1125	0.91
Chena (Alaska)	5015	14	95	0.54
Gunnisson (Colo.)	21,000	15	81	0.5
Gila (N. Mex.)	46,600	29	20	0.24
Strike-a-light Creek (NSW)	216	8	90	0.23
x		11.9		2.50

Source: modified after Douglas (1969) Table III

will be low in very cold areas. The reasons given for such a view include the paucity of true soil development, the presence of 'fresh' rock outcrops, the purportedly low rates of chemical reactions associated with low temperatures, and the paucity of biomass to provide organic weathering agents. However, the bulk of the meagre store of empirical data relating to such areas tends to indicate that this is an oversimplification and an exaggeration. Moreover, there are some counter arguments that can be used to suggest that rates should not be very low in cold areas. Reynolds and Johnson (1972), for instance, who found very high rates of chemical denudation in the high altitude, high relief Northern Cascade Mountains of North America, argued that glaciation and mass wasting would constantly produce and renew fresh rock surfaces, which could then be attacked by rainwater. They also suggested that in such an

Table 2.13 Dissolved silica budgets for selected catchments

Location	Input (t km^{-2} a^{-1})	Output (t km^{-2} a^{-1})	Net loss or gain (t km^{-2}a^{-1})	Annual precipitation (mm)	Annual runoff (mm)
Jamieson Creek (BC, Canada)	0.075	9.20	−9.13	4540	3670
*Kali Mondo (Indonesia)	< 1.06	53.8	−52.7	4668	3590
*Behana Creek (Queensland)	–	27.9		4000	2325
H. J. Andrews (Oregon, USA)	trace	11.36	−11.35	2370	1545
*Ei Creek (Papua, New Guinea)	0	28.8	−28.8	2700	1480
*Davies Creek (Queensland)	–	18.0		1400	900
*Sungai Gombak (Malaya)	–	12.8		2360	856
Hubbard Brook (NH, USA)	0	2.38	−2.38	1322	833
*Rio Tanama (Puerto Rico)	–	16.5		2000	630
East Twin (Somerset, UK)	0.52	2.74	−2.22	1078	514
Haartsbach (Luxemburg)	0.05	2.75	−2.70	774	383
*Kaingaroa (NI, New Zealand)	–	4.90		1500	170
*Amitioro (Ivory Coast)	1.31	1.03	0.23	1323	120

* Tropical and sub-tropical examples.
Source: modified after Bruijnzeel (1983) table 2.

environment there would be a very high surface area of rock exposed to water, because of the presence of rock flour, talus, moraine, fractured cliff faces, and rock rubble on ridges and summits. Moreover, some Arctic environments may have large amounts of precipitation, which is rendered all the more effective because of the limited evapotranspirational losses, and this is seen as a probable explanation for the high rates of chemical denudation encountered in a maritime galcierized catchment in Canada (Eyles et al., 1982). Not only may rates of chemical denudation be high in such areas, they may also be more important than other types of denudational process, as Rapp (1960) found in Arctic catchments in Scandinavia.

Work in the Alpine environment of Colorado, USA, also tends to demonstrate the significance of solutional denudation in cold and nival environments. Indeed, according to Caine (1992, p. 64) in the Colorado Front Range, 'Studies all point to solute removal as a dominant influence in contemporary landscape modification.' In a study of the Green Lakes Valley, an unglacierized catchment, solute yields exceeded by an order of magnitude rates of suspended sediment transport in the basin. Another important finding was that within the catchment solute loads were highest in subcatchments where winter snow accumulation is greatest. The importance of solutional denudation in comparison with some other processes has also been illustrated in the Canadian Rockies (Smith, 1992), where in the Mount Rae area, solute transport achieves 99.2 per cent of geomorphic work. Other processes thought to be 'typical' of such alpine environments, including solifluction, rockfalls and avalanches, achieved much less (table 2.14). Another confirmatory study is that of Lewkowicz (1988) on Banks Island,

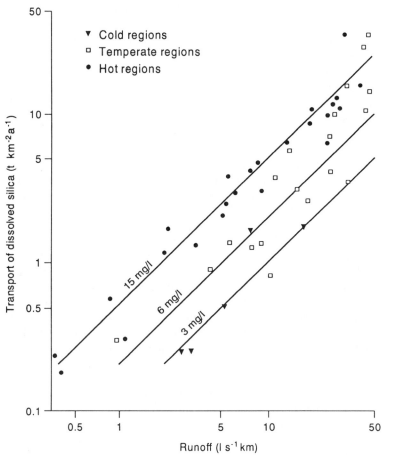

Figure 2.12
The rate of transport of dissolved silica plotted against runoff (specific discharge) for cold, temperate and hot regions (source: Meybeck, 1987).

Table 2.14 Estimates of geomorphological work in the Mount Rae area of the Canadian Rockies (after Smith, 1992, table 9.6)

Process set	Geomorphological work	
	Amount (10^6 J km^2 a^{-1})	Percentage of total work
Ground squirrel burrowing	0.07	0.01
Solifluction	1.69	0.2
Rockfalls and avalanches	4.81	1.6
Solute transport	860.60	99.2

northern Canada. He found (p. 355) that 'Solute removal was far more effective than suspended sediment erosion, and denudation was 8–30 times greater for this process' (2.0–74.1 mm 1000 a^{-1}, compared with 0.4–2.6 mm 1000 a^{-1}).

On the other hand, some studies of the way in which rates of solute removal vary with altitude show that rates can be lower in high, alpine areas in comparison with lower altitudes. Drever and Zobrist (1992), for example, studied the solute loads of streams in the southern Swiss Alps over a range of altitudes from 220 m to 2400 m. They found that rates decreased by a factor of 25 from lower elevations to higher. They attributed this partly to the role of diminishing temperature, but also believed that the presence of thin soils and limited vegetation cover at higher altitudes were the prime causes of this trend.

Within cold, high altitude situations, the rate of weathering may vary according to local micro-climatic conditions, and particularly with respect to such factors as snow cover. This was brought out by Benedict (1993) who used measurements of phenocryst relief on granodiorite outcrops in the Colorado Front Range which had been exposed by late Pleistocene deglaciation. Rates of bedrock surface lowering by all weathering processes ranged from 0.0001 to 0.0013 mm a^{-1}. Weathering appeared to be least effective beneath late-lying snow banks that protect the rock from freeze–thaw and wetting and drying cycles for most of the year, and in those sites that are snow-free in winter, where dryness and the binding effects of crustose lichens discourage granular disintegration. Weathering was seen to be most effective in areas of thin to moderate snow cover where meltout occurs early, moisture is available at critical seasons, and episodic lichen 'snowkill' ensures that bare rock is exposed.

Meybeck (1987) has gone some way towards quantifying the lithological control on rates of denudation that was perceived by Walling and Webb to be one of the prime reasons why there was such a weak relationship between climate and rate of denudation at the zonal scale. As figure 2.13 suggests, drainage basins underlain by metamorphic and plutonic rocks tend to have lower values of chemical transport for a given runoff level than do volcanic rocks. The highest rates occur for sedimentary rocks. He also established a relative erosion rate normalized to granite weathering:

relative erosion =
$$\frac{\text{(\% chemical weathering / \% outcrop) for given rock type}}{\text{(\% chemical weathering / \% outcrop) granite}}$$

The relative rates were as follows:

Granite 1
Gneiss and mica schist 1
Gabbro 1.3
Sandstone 1.3

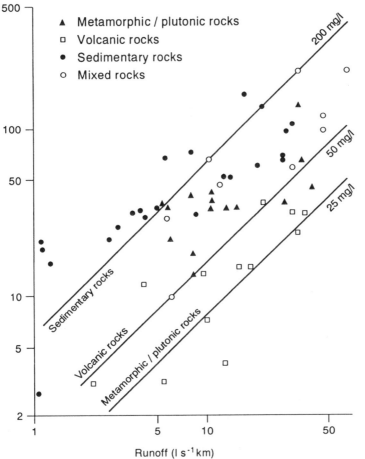

Figure 2.13
The rate of transport of all major ions plus dissolved silica plotted against runoff (specific discharge) for various major drainage basins underlain by sedimentary, volcanic, and metamorphic and plutonic rocks (source: Meybeck, 1987).

Volcanics 1.5
Shales 2.5
Miscellaneous metamorphics (including serpentines, marbles
    and amphibolites) 5
Carbonates 12
Gypsum 40
Rock salt 80

Although river water analyses are most usually employed to establish rates of solutional denudation, they can also be used to estimate the time required to convert fresh rock into a particular type of weathered material. This is a matter that was discussed earlier in the context of duricrust formation. A good example of an attempt to estimate the time required to convert a one metre thickness of fresh rock into kaolinite is provided by Nahon (1991). Using data presented in table 2.15 he demonstrated that for a range of igneous and metamorphic rock types under a range of

Table 2.15   The time required to weather 1m of rock to kaolinite under different climate conditions

Location and lithology	Average annual precipitation (mm)	Average SiO$_2$ content of water (mg l^{-1})	Number of years for transformation
Granite, Norway	1250	3	85,000
Granite, E. France	850	9.2	52,000
Granite S. France	680	11.5	41,000
Migmatite, S. France	540	5.9	100,000
Migmatite, Ivory Coast	540	20	65,000
Amphibolite, S. France	640	14	68,000
Basalt, Madagascar	1500	16	40,000

Source: modified after Nahon (1991).

Table 2.16   Mean lifetime of one mm of fresh rock

Rock	Climate	Lifetime (years)
Acid	Tropical semi-arid	65–200
	Tropical humid	20–70
	Temperate humid	41–250
	Cold humid	35
Metamorphic	Temperate humid	33
Basic	Temperate humid	68
	Tropical humid	40
Ultrabasic	Tropical humid	21–35

Source: modified after Nahon (1991).

annual rainfall conditions such conversion requires between 40,000 and 100,000 years.

Nahon (1991) also used river water analyses to estimate 'the mean lifetime' of one mm of fresh rock (table 2.16). The prime difference that emerged in this work was not that tropical conditions were especially conducive to rapid rock decomposition, but that there was a general tendency for basic and ultrabasic rocks to weather about two and a half times faster than acid rocks.

Although most attention tends to be devoted to the study of the inorganic load of rivers, as represented by ions like silica or calcium, some rivers also carry an appreciable load of organic material derived from the land. One measure of organic load is dissolved organic carbon (DOC). DOC loads in rivers per unit area correlate significantly with runoff (figure 2.14), but there may also be a significant climatic zone control (Spitzy and Leenheer, 1991). Concentrations range from less than 1 mg l^{-1} in alpine streams to more than 20 mg l^{-1} in some tropical rivers and rivers draining swamps, wetlands or polluted regions (table 2.17). Actual loads also tend to be relatively high for tropical rivers, and tropical rivers probably carry around 68 per cent of the DOC that is taken to the oceans. The DOC load of the Amazon is 3.2 t km^{-2} a^{-1}, and that for the Orinoco is 6.1.

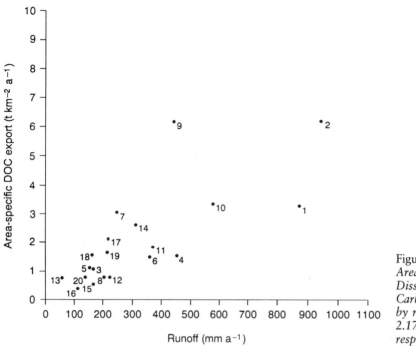

Figure 2.14
Area-specific annual Dissolved Organic Carbon (DOC) export by rivers given in table 2.17 versus their respective runoff.

Rates of karst denudation

In the mid-1950s Corbel, a French geomorphologist, began to make a systematic study of the rates of karst denudation based on a spatially extensive programme of water sampling. The formula that he employed was (Corbel, 1959):

$$\text{Limestone denudation rate in } m^3 \ km^{-2} \ a^{-1} = \frac{2.5 \ E.t.n.}{1000}$$

where E is the runoff in decimetres, t is the C_aCO_3 concentration in mg l^{-1}, and 1/n is the fraction of the catchment formed of a carbonate lithology. A bulk density of 2.5 gm cm^{-3} was assumed. Corbel's sampling programme suffered from certain deficiencies. In particular he often estimated discharge from climatic data (precipitation–evaporation) rather than from discharge records themselves, and he also generally took too few samples to be able to obtain a reliable mean carbonate concentration. Nonetheless it was a highly stimulating piece of pioneering work that led both to improvements in technique and to attempts to test the validity of one of Corbel's main contentions – namely that rates of limestone denudation tended to be higher in cooler regions than in the humid tropics. This was contrary to the normal stance at the time.

Corbel (1964) was certain of the importance and vigour of limestone dissolution in cold environments (p. 399) (my translation):

Table 2.17 Dissolved organic carbon transport for some major rivers

River	Discharge (km³ a⁻¹)	Area (10³ km²)	Runoff (mm a⁻¹)	DOC concentration range (mg l⁻¹)	(t km² a⁻¹)
1. Amazon	5520	6300	876	3–5	3.2
2. Orinoco	1135	950	996	2–5	6.1
3. Parana	480	2800	171	–20	1
4. Uruguay	158	350	451	2–8	1.4
5. Mississippi	439	3267	154	2–8	1.1
6. St Lawrence	413	1150	359	3–5	1.4
7. Yukon	210	840	250		3.0
8. Columbia	135	670	202		0.7
9. Yangtsekiang	883	1950	453	5–23	6.1
10. Brahmaputra	609	580	1050	1–6	3.3
11. Ganges	366	975	375	1–9	1.7
12. Indus	211	950	222	2–22	0.8
13. Huanghe	44	745	59	5–25	0.72
14. Zaire	1237	4000	309	3–10	2.6
15. Niger	152	1125	171	2–6	0.5
16. Gambia	4.6	42	110	1–4	0.27
17. Lena	533	2430	219		2.1
18. Ob	419	2550	164		1.5
19. Yenisei	562	2580	218		1.6
20. Mackenzie	249	1810	138	3–6	0.7
Sum	13759.6				

Source: from Spitzy and Leenheer (1991) table 9.2.

Carbon dioxide (and thus limestone) is more soluble under cold conditions and furthermore in a cold climate evaporation is much less strong so that runoff is more important. It follows that for comparable levels of precipitation and relief, the solution of limestones is ten times more strong in cold, snowy regions than under hot climates. For example, for regions receiving from 1000 to 1600 mm of precipitation, denudation of limestones varies from 160 m³ km⁻² a⁻¹ in a cold region to 16 in a hot region.

The solubility of CO_2 in water, which determines the solubility of limestones, is inversely proportional to temperature, doubling as water is cooled from 20°C to 0°C. It was this that led Corbel to make this proposal that limestone karst landforms develop most rapidly in arctic and alpine regions.

By the mid-1970s enough studies had been done, based both on relatively reliable long-term discharge records and on an adequate water chemistry sampling programme, to enable Smith and Atkinson (1976) to attempt a further review of the climatic patterning of rates of karstic denudation. Their review of available data is summarized in figure 2.15. Their findings were of great interest. The most important of these was that rates of limestone solutional removal increased with increasing runoff. A secondary finding was that rates were less in areas of bare rock and greater in soil and vegetation covered terrain. The data gave little evidence that tem-

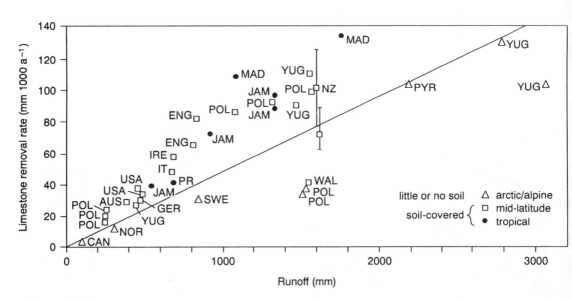

Figure 2.15
A plot of limestone removal rate in mm 1000 a⁻¹ against runoff (source: Jennings, 1985, fig. 67): AUS – Australia, CAN – Canada, ENG – England, GER – Germany, IRE – Ireland, IT – Italy, JAM – Jamaica, MAD – Madagascar, NZ – New Zealand, NOR – Norway, POL – Poland, PR – Puerto Rico, PYR, Pyrenees, SWE – Sweden, USA – United States, WAL – Wales, YUG – Former Yugoslavia.

perature was a very significant factor except in so far as it affected evapotranspiration loss of water in the system. The key factor, therefore, appears to be firstly water availability, and secondly the greater amounts of aggressive soil CO_2 in soil and vegetation mantled areas.

The amount of CO_2 can be greatly increased in the pores of a soil because it is discharged there by plant roots and bacteria. Partial pressures of CO_2 of 3 per cent, compared with 0.03 per cent in the ordinary atmosphere, are of frequent occurrence. Thus water passing through a soil will become enriched in this gas and so be capable of dissolving much more limestone than can water at the same temperature that has only encountered the ordinary atmosphere. Furthermore, it can be argued (see, for example, Ford, 1993) that this soil and vegetation factor works in opposition to the inverse relationship between CO_2 solubility and temperature in that the amount of soil CO_2 production tends to increase with temperature.

Smith and Atkinson's data have subsequently gained acceptance and amplification from Jennings (1985, fig. 67) and, in the context of the Mediterranean lands (plate 2.5) from Maire (1981). Maire draws a distinction between four types of climatic environment in the Mediterranean karstlands:

Plate 2.5
In the Mediterranean basin, karst landforms, such as these karren developed on Burdigalia limestones at Lluch in Mallorca, demonstrate the power of limestone solutional processes. The prime control of rates appears to be an area's amount of precipitation.

The high sub-arid karsts with a denudation rate of 20 to 40 mm 1000 a^{-1}

The high semi-humid karsts with a rate of 40–60 mm 1000 a^{-1}

The high humid karsts with rates of 60–90 mm 1000 a^{-1}

The hyperhumid high karsts with rates in excess of 90 mm 1000 a^{-1}.

His data are summarized in figure 2.16.

The importance of precipitation amount in determining rates of solutional denudation in carbonate rocks has been confirmed by studies of the rates at which marble tombstones have been weathered over time. Neil (1989) investigated Carrara marble tombstone sites in eastern Australia over a substantial rainfall range (531 to 2224 mm a^{-1}), and found a strong correlation between gross rainfall amount and weathering rate (figure 2.17). In addition he found that proximity to the sea was a major control, and while this is mainly due to a concomitant decline in rainfall as one moves inland, there is also a decline at constant rainfall, suggest-

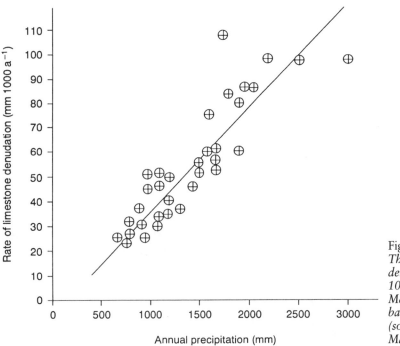

Figure 2.16
The rate of limestone denudation in mm 1000 a^{-1} in the Mediterranean basin based on 34 studies (source: modified from Maire, 1981, fig. 3).

ing that coastal rainfall, with its high chloride content, is more aggressive towards limestone surfaces.

Current rates of solutional denudation in Britain

The use of karst drainage water chemistry data and discharge determination enables an estimation of the rate at which karstic denudation is currently occurring in Britain. Rates, which conventionally are expressed as $m^3 km^{-2} a^{-1}$, are listed in table 2.18. The data can conveniently be divided into three main groups, based on the age of the limestone from which they are derived: Carboniferous, Jurassic and Cretaceous (Chalk). What emerges form this is that within Britain there is a considerable range in values, even within areas with broadly similar climatic conditions. The lowest values have been reported from South Wales (16 mm 1000 a^{-1}) whereas values of over 100 mm 1000 a^{-1} have been reported from the Mendips (in south-west England) and from the island site of Lismore in western Scotland.

There is, however, no great difference observed in mean values between the three main rock classes: Carboniferous is 55 mm 1000 a^{-1}, Jurassic is 59 mm 1000 a^{-1}, and the Chalk is 50 mm

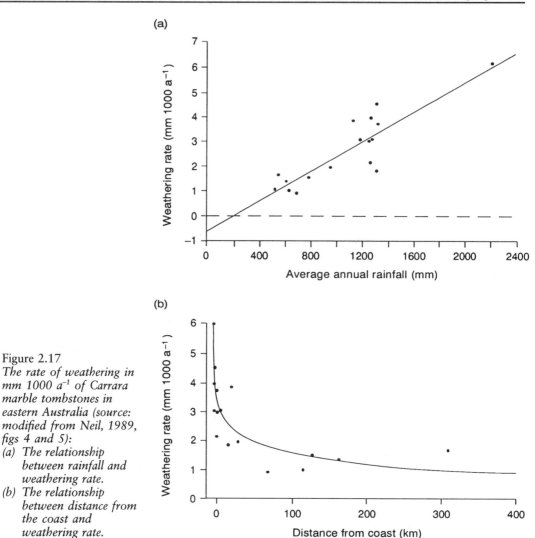

Figure 2.17
The rate of weathering in mm 1000 a⁻¹ of Carrara marble tombstones in eastern Australia (source: modified from Neil, 1989, figs 4 and 5):
(a) The relationship between rainfall and weathering rate.
(b) The relationship between distance from the coast and weathering rate.

1000 a⁻¹. This suggests that the very different forms of karst encountered on the three rock classes are not the result of very different rates of solutional denudation. They also indicate that rates of solutional denudation in Britain are broadly comparable to those found in other climatic regimes (see Smith and Atkinson, 1976). The rates also indicate that substantial landscape modification may have occurred even within the span of the Pleistocene. A rate of denudation of 50 mm 1000 a⁻¹, if maintained during the vicissitudes of climatic change, is equivalent to a rate of land surface lowering of around 80 m during the 1.6 million years of the Pleistocene. It is perhaps not surprising, given such rates as

Table 2.18 Rates of limestone solution in England and Wales

Location	Rock type	Rate (mm 1000 a⁻¹)
North Cotswolds	Jurassic	35
Corallian Scarp, Oxfordshire	Jurassic	74
North Oxford Heights	Jurassic	60–66
Berkshire Downs	Chalk	67
East Anglia	Chalk	25
Dorset	Chalk	50
South Downs	Chalk	55–65
E. Yorkshire	Chalk	39
South Wales	Carboniferous	16
Mendips (E)	Carboniferous	50–102
Mendips (S. Central)	Carboniferous	81
Mendips (Cheddar)	Carboniferous	38–45
Mendips (Cheddar)	Carboniferous	23–29
Gower	Carboniferous	30
Derbyshire	Carboniferous	75–83
NW Yorkshire	Carboniferous	83

Source: from various sources in Goudie (1990).

these, that the Cretaceous Chalk has been removed over large tracts of the British Isles during the course of the Tertiary.

Rates of non-karstic chemical denudation in the UK

Walling and Webb (1981) have undertaken a detailed survey of the amount of dissolved material currently being transported in British rivers. They have found that the highest concentrations of dissolved material (as expressed by specific conductance values) occur in lowland areas underlain by less resistant rock types (especially various types of sedimentary rock). By contrast, to the west of the Tees-Exe line, where rocks are more resistant and where high runoff from moist uplands causes a dilution effect, specific conductance levels are low (figure 2.18).

They also looked at the pattern of total dissolved solids loadings expressed as the rate of removal in t $km^{-2} a^{-1}$ (figure 2.19). This is based on the analysis of solute concentrations and runoff volumes. Values range between 10 and 400 t $km^{-2} a^{-1}$. Areas in the east of England, being drier, have low or medium loadings as do upland areas with highly resistant rocks. If one makes allowance for the non-denudational component of the dissolved loadings, it is possible to calculate a rate of chemical denudation (figure 2.20). This demonstrates that in general rates of chemical denudation for non-carbonate rocks are less than 40 $m^3 km^{-2} a^{-1}$ (Walling and Webb, 1986), and that more than 50 per cent of the country experiences an annual rate which is less than 20 mm 1000 a^{-1}. Rates on Silurian greywackes in mid-Wales are even lower than this – only around 2 mm 1000 a^{-1} (Oxley, 1974) – and comparably low values have

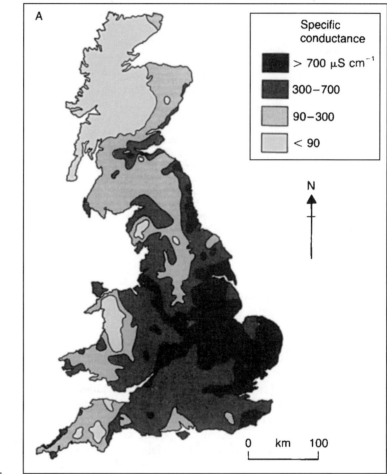

Figure 2.18
Mean background loads
of specific conductance in
British streams based
primarily on data
collected in the period
1977–9 (source: Walling
and Webb, 1981, fig. 5.3).

been found for the Old Red Sandstone core of the Mendip Hills
(Waylen, 1979) (table 2.19).

Conclusions

With respect to weathering phenomena we have presented data
on, *inter alia*, rates of duricrust and tufa formation and hardening
and the formation of deep weathering profiles. Under favourable
conditions deep weathering rates may be higher than often supposed and deep profiles may have formed in the Quaternary rather
than over more extended time-scales. Many weathering phenomena, including rinds, seems to show a decelerating rate of development with time. The role of biological processes in affecting

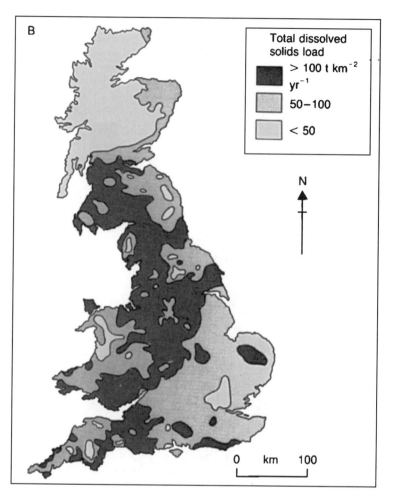

Figure 2.19
*The pattern of annual
total dissolved solids loads
in British streams (source:
Walling and Webb, 1981,
fig. 5.9).*

weathering rates needs much greater attention, but the role of air
pollution in causing accelerating rates of rock decay in cities appears clear.

The most used method to determine rates of chemical denudation has been the study of stream quality and discharge. The non-
denudational component of river loads is, however, important and
on a global basis amounts to perhaps 40 per cent of the total load.
It is especially important in coastal areas and in those locations
where dissolved loads are low. On a global basis stream dissolved
loads average just under 40 t km^{-2} a^{-1} but range from around zero
to over 300 t km^{-2} a^{-1}. Many factors determine such rates, including rock type and climate, but rates of chemical denudation
appear on a global basis to be very much less than rates of mechanical denudation (as represented by stream suspended loads).

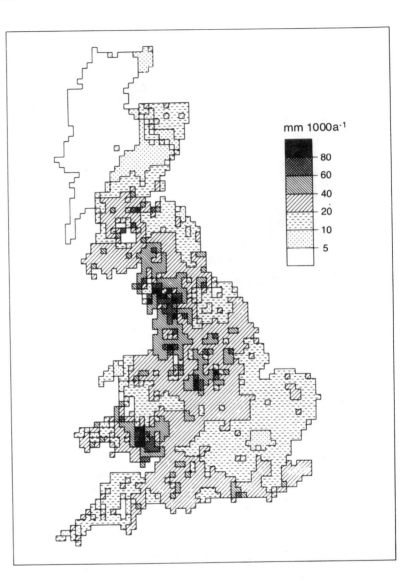

mm 1000a⁻¹

Figure 2.20
Estimated rates of chemical denudation in mm 1000 a^{-1} in Britain (source: Walling and Webb, 1986, fig. 7.18).

High rates of chemical denudation occur in mountainous areas with high rainfall and the lowest rates occur in dry environments. Controversy attaches to the question of whether rates are especially high in the humid tropics and in particular as to whether silica loadings and concentrations are exceptionally high there. Equally, although it has often been surmised that rates will tend to be lower in cold environments, some recent studies have tended to show that such a view may be misfounded. Less controversy surrounds the question of the way in which rock type influences rates of chemical denudation, and the highest rates occur in sedimentary rocks, especially carbonate rocks and evaporites. How-

Table 2.19 Rates of solutional denudation on non-carbonate rocks in Britain

Lithology	Location	Solutional denudation (mm 1000 a^{-1})
Old red sandstone	Mendip	1.6 Waylen (1979)
Clay-with-flints	SE Devon	14.0 Walling (1971)
Keuper Marl	SE Devon	22.8 Walling (1971)
Upper Greensand	SE Devon	16.8 Walling (1971)
Keuper Marl-Upper Greensand	SE Devon	33.8 Walling (1971)
Keuper Marl-Upper Greensand	SE Devon	27.6 Walling (1971)
Silurian Greywackes	Montgomeryshire	2.1 Oxley (1974)
Silurian Greywackes	Montgomeryshire	1.6 Oxley (1974)
Middle Jurassic-Liassic clays and shales	North York Moors	20.2 Imeson (1974)
Granite	Dartmoor	5.0 Williams et al. (1986)
Sandstones and shales	Midlands	10.5 Foster and Grieve (1984)
Lower Old Red Sandstone	Fife	19.0 Tricker and Scott (1980)
Upper Old Red Sandstone	Fife	43.0 Tricker and Scott (1980)

ever, rates of chemical denudation in limestone areas have also created controversy with respect to the role of temperature, but the key factor appears to be water availability and the nature of the soil and vegetation cover.

THREE

The fluvial environment

Introduction

Although on a global basis the quantity of water in rivers is only about 0.0001 per cent of the total, rivers are fundamental to an understanding of the geomorphology of most parts of the earth's surface. River channels are far and away the most dynamic components of the landscape and rivers are the conduits by which water and sediment are transferred from slopes to oceans, seas and lakes. In the previous chapter we saw how rivers were conduits for the chemical materials being evacuated from catchments. In this chapter we start by looking at more general issues of fluvial denudation, including the rate at which suspended material (kept in suspension by the turbulence of the flowing water) and bed load (composed of coarser particles that are moved on or near the bed of the stream) are transported in fluvial systems.

Techniques for determining rates of fluvial denudation

To estimate rates of fluvial denudation over relatively short time-spans, geomorphologists have traditionally used two main types of information: that provided by river sediment loads, and that provided by sedimentation in closed basins (e.g. lakes and reservoirs).

The most widespread method to estimate rates of erosion is to use river discharge and sediment load data. Such data are available for many hundreds of catchments on a global basis (Walling and Kleo, 1979), though the reliability, frequency, regularity and length of the record varies greatly. In addition to determining volumes of discharge passing down a particular river or stream, it is also necessary to calculate the concentrations of the suspended and bed loads, and each of these has its problems.

Suspended sediment sampling can be achieved with four different types of equipment (Ward, 1984):

1. *Manual withdrawal of single samples* involves the observer dipping a wide-mouthed bottle into the flow. Inaccuracies can result if samples are obtained from the bank and only from the surface, but the method is quick and simple.
2. *Mechanical withdrawal of single samples* involves the use of pumps to withdraw samples from the river automatically at known times, and to place them in bottles for subsequent analysis.
3. *Multiple sampling across the section* uses a sample bottle that is encased in a streamlined housing, with the intake nozzle extending forwards into the flow from the front of the housing. The samplers are lowered from a bridge, boat, cable, or by wading so that many positions in the vertical and horizontal are sampled. These instruments are useful for measurements in large rivers, where there is a substantial sand load and where changes of stage do not occur too rapidly.
4. *Direct readings from in situ instruments* employ permanently installed instruments in the river which rely upon the principle of light or gamma ray attenuation.

Suspended sediment concentrations in rivers undergo marked and rapid fluctuations, and although there are some items of equipment which permit continuous monitoring of sediment concentration data (e.g. nuclear probes, turbidity meters and automatic pump samplers) for many rivers the record is imperfect. However, continuous records of stream discharge are more readily available at most measuring stations.

In such situations, to estimate sediment yields over long periods indirect load calculation procedures, involving either interpolation or extrapolation of the available concentration data, must be used (Walling and Webb, 1981). Interpolation procedures assume that the values of concentration or sediment discharge obtained from instantaneous samples are representative of a much longer period of time. Extrapolation procedures (the rating curve technique) involve extrapolating a limited number of sediment concentration measurements over the period of interest by developing a relationship between concentration or sediment discharge and stream discharge, and by applying this relationship to the streamflow record.

Estimates of suspended sediment load based on these interpolation and extrapolation techniques may involve considerable errors and uncertainties, and the literature contains many examples of major discrepancies presented by different scientists working on the same rivers. These result because of differences in sampling strategy, data availability and load calculation procedures. A classic example of this comes from the Cleddau River in New Zealand where, using different rating curve techniques on the same basic discharge and sediment concentration data, Griffiths (1979) and

Adams (1980) calculated loads that were different by nearly two orders of magnitude (275 t km^{-2} a^{-1} and 13300 t km^{-2} a^{-1} respectively). In their investigation of the River Creedy in Devon, England, Walling and Webb (1981) found that certain interpolation procedures underestimated the true loads by 70 per cent or more, whereas certain extrapolation techniques underestimated sediment loads by between 83 and 23 per cent.

Bed load is one of the most difficult components of river load to measure with any degree of accuracy. The following approaches are available (Ward, 1984):

1. *Channel and bed material measurements followed by calculation*. This method enables the bed material load in an alluvial reach of a sand bed river to be predicted, using equations developed by Einstein (1950). Measurements are made of a cross-sectional area of the channel, mean gradient, grain size characteristics of bed material, etc. in a test reach close to a flow gauging station. There is no requirement that measurements be made during the flood for which information is required and the method has the advantage of being applicable to rivers which change their stage rapidly. However, there are errors and uncertainties in some of the coefficients employed, so that the method may have limited accuracy in certain circumstances.

2. *Bed-load samplers* are traps that are lowered to the bed during times of significant flood flows and allowed to collect a sample of the bed load. Ward identifies three major problems that prevent good performance: the disturbance of the flow and hence the pattern of bed-load movement by the sampler; the fluctuating nature of the bed-load transport; and improper operation whereby the sampler digs into the bed layer. Much used samplers include the Arnhem Sampler and the Helley-Smith Sampler (Emmett, 1984).

3. *Slot traps* are traps set in the beds of small rivers, which collect bed load as it moves down the river. The material that is trapped can be removed mechanically (i.e. by bulldozer) or by sophisticated systems employing hydraulically operated gates and conveyor belts.

4. *Tracers* enable bed material movements to be monitored. Artificial labelling techniques can be employed using paints, fluorescent dyes, and radioactive elements.

Calculating overall rates of transport from such techniques faces the same sorts of problems that have already been discussed for the suspended sediment load.

In assessing rates of net denudation, allowance needs to be made for inputs, whether as solutes in precipitation (see chapter 2) or as dry deposition. It has become increasingly clear in recent years that substantial quantities of primarily silt and clay sized material may be blown into catchments by dust storms that have transported material from susceptible deflational surfaces, especially in peridesert and periglacial areas (Goudie, 1983). Such rates of input may be of the same order as rates of sediment output, being characteristically of the order of 50 t km^{-2} a^{-1}. Techniques for measuring dry deposition are discussed by Steen (1979).

Lake and reservoir sedimentation

Given the uncertainties attached to the use of sediment load and discharge data, there have been many attempts to use the volumes of sediment that have accumulated in lakes and reservoirs over a known period of time (see Trimble and Carey, 1992, for a recent review).

Reservoir sedimentation surveys have various advantages over streamflow and suspended sediment sampling methods for estimating sediment yields for catchments. In particular, the sediment yield for past years can be determined with just one survey, thereby obviating the need to wait years for the collection of streamflow data. As Trimble and Bube (1990) point out, sediment in a flowing stream must be sampled over a long period of time in order that all catchment and channel conditions have been experienced. Sampling omission during a large event may mean a substantial error in the annual average. The sedimentation survey can also be made at a convenient time and once the necessary ranges and benchmarks are established, a reservoir can be resurveyed in only a few days (Rausch and Heinmann, 1984).

There are, however, some problems with this approach. It may be difficult to define the original reservoir bottom topography, and one also needs to know the history of the reservoir, when it started to fill and whether during its history any sediment has been removed. Certain assumptions also have to be made about the quantity of material discharged through the spillways (i.e. the trap efficiency). The original bottom of the reservoir needs to be established (e.g. by coring) and then sediment volumes established by various survey techniques. The sediment also needs to be sampled with such devices as piston-type samplers or gamma density probes; and in complex and varied sediments this may require extensive sampling. Once the volume and density of the sediment are determined then in order to determine the average annual sediment yield from the contributing catchment the weight of deposited sediment must be adjusted for reservoir sediment trap efficiency.

Reservoir-sediment trap efficiency depends on a variety of factors: the nature of the inflowing water and sediment, the reservoir storage dimensions and properties, and the nature of the discharge from the reservoir. These factors are highly variable from reservoir to reservoir and are the subject of a voluminous literature (Heinemann, 1984).

One further consideration is that to assess the true sediment yield from the contributing catchment it is necessary to make allowances for factors such as the organic deposits in the reservoir sediments. This will include both soft tissues and organic carbon as well as diatoms (McManus and Duck, 1985). Atmospheric dust inputs and bank erosion could also contribute to the total sediment mass, the latter as a result of fluctuating water-levels, wave action, frost action, heaving by lake ice and physical disturbance such as trampling by cattle (Foster et al., 1985).

The study of rates of sedimentation in natural lakes can also give an indication of rates of catchment erosion, though for this it is essential to be able to date the sediments with some degree of accuracy. In some cases the age of a lake basin may be known because it was created at a particular moment (e.g. a river dammed by a landslide that occurred at a known date – see, for example, Eardley, 1966). Alternatively it is necessary to use a range of dating techniques, which include varve chronologies, radio-carbon dating, Lead-210 and Caesium-137 dating, and a palaeomagnetic time-scale.

The sediment thicknesses found between particular lake sediment horizons or layers can also provide an estimate of sediment yields, at different time intervals, though to be successful an adequate density of coring points is required. This in turn requires that one should use rapid methods for correlation of synchronous levels if the quantity of laboratory work is not to become unwieldy. Such methods include varves, abrupt changes in colour of sediment, visible horizons produced by volcanic ash inputs, micro- or macro-fossil remains, and zone boundaries in pollen or diatoms. Increasing use is also being made of magnetic properties (Oldfield, 1981).

Table 3.1 provides some data on rates of sedimentation in selected lakes (from Kukal, 1990, Table 36). Lakes which are filled with coarser clastic material display faster rates of sedimentation than those in which only clays or biochemical precipitates accumulate. Sedimentation is also faster in smaller lakes compared with larger, and in those where accelerated erosion in their contributing catchment areas has resulted from the human impact. In addition, accelerated organic accumulation may occur as a result of 'cultural eutrophication' resulting from the addition of large quantities of nutrients (e.g. nitrates and phosphates) derived from human activities.

According to Kukal (1990, p. 99) the mean rate of sedimentation in lakes can be estimated as being 0.3 cm a^{-1}.

Gross denudation techniques

As already mentioned, the calculation of overall rates of denudation can be approached in one of two ways: either by using an

Table 3.1 Characteristic rates of lake sedimentation

Lake	Type of lake and sediment	Rate of sedimentation [cm ka⁻¹]
Great Lakes, N. America	glacial; muds, varves	15
Manitoba	glacial oligotrophic; muds	50–300
Chicot, Arkansas	oligotrophic; muds and sands	1800
Michigan	glacial oligotrophic; calcareous varved sediments	300
Maxinkuckee	eutrophic; lacustrine marl	30
Ontario	glacial; varves	20–110
Ladoga	oligotrophic; muds	610
Onega	oligotrophic; muds	710
Loch Lomond	glacial; varves and muds	100
Swedish Lakes	eutrophic; gyttja	100–200
Olof Jone Damm	dystrophic; peat	530
Vierwaldstädt	alpine oligotrophic; calcareous muds	1040–3170
Wallen	alpine oligotrophic; calcareous muds	1130
Brienz	alpine oligotrophic; calcareous muds	1790
Zürich	glacial oligotrophic; muds, calcareous varves	40–70
Geneva	oligotrophic; sands	1790
	muds	80–230
North German Lakes	oligotrophic; eutrophic; lacustrine marl	100–300
Bavarian Lakes	oligotrophic; muds	200
Baldeney (Ruhr)	subsiding due to undermining; muds and sand	1000

Source: Kukal (1990) table 36.

essentially geological approach, or by measuring present-day operation of processes. The former class of techniques enables calculation of rates for relatively large areas over relatively long time-spans.

For example, the basis of Ruxton and McDougall's (1967) approach to estimating rates of denudation in Papua was that if a dissected volcanic landscape possessed earlier or 'original' forms that could be reconstructed and dated by potassium argon, then by measuring the volume of material removed, denudation rates could be calculated. Measurement of the volume of material removed was achieved by using generalized contours to reconstruct the original surface of a strato-volcano, and then estimating the amount of ground lowering of concentric sectors distant from the centre by measuring the difference between present-day and the original cross-sectional areas.

Clark and Jäger (1969) used isotopic dates on biotites in the European Alps to infer long-term rates of denudation. They suggested for their model that the apparent ages of biotites from rocks metamorphosed during the Alpine orogeny represent the time at which the biotites had cooled to the point where loss of radiogenic elements no longer occurred. By calculating the required degree of cooling, and estimating therefrom the depths the rocks were at a given date in the past, they were able to calculate rates of denudation. Similarly, in their study of Mesozoic denudation rates in New England, Doherty and Lyons (1980) used the apparent discrepancies between potassium argon dates and fission track dates on zircon and apatite for the White Mountain Plutonic-Volcanic Series. The basis of this method is that the K-Ar dates give an initial date for the emplacement of the intrusives and that apatite and zircon have a distinctive temperature range at which tracks will begin to accumulate during cooling. The assumed temperatures are 200°C for zircon and 100°C for apatite. Thus the differences in the three determinations are temperature related and the temperature is in turn related to depths of emplacement. Assuming a 25 to 30°C km^{-1} geothermal gradient, it is possible to calculate thereby depths at different times in the past and from that to calculate the thickness of overlying rock that has been removed. Such techniques have also been employed in Britain by Lewis et al. (1992), who thereby estimated that erosion in the Tertiary had removed some 3 km of overburden over much of north-west England. The basis of these methods is shown in figure 3.1.

The use of sediment volumes of known age on continental shelves was the technique favoured by Gilluly (1964) to calculate erosion rates over the long term in the eastern USA. By planimetry the area and volume of Triassic and younger sedimentary rocks was calculated. Adjustments were made for the difference in density between source rocks and the deposits derived from them. It was believed, largely because of the absence of calcareous rocks, that the sediments were supplied primarily as suspended and bed-load material. Problems were encountered in establishing the source area of the sediments because of tectonic changes, river capture, etc. since the Triassic. Another problem was presented by the re-working of coastal plain sediments, which meant that some of the measured volume of sediment had been re-worked one or more times during its history.

Bishop (1985) used a combination of methods in south-east Australia to estimate the late Mesozoic and Cainozoic denudation rates:

1. Average rates of post-early Miocene incision were determined by dividing the height difference between the present channel of the Lachlan River and the top of the adjacent basalt remnants (age, 20–21 Ma) which themselves mark

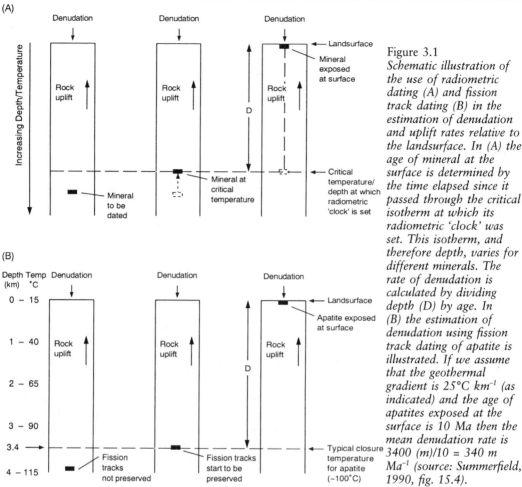

Figure 3.1
Schematic illustration of the use of radiometric dating (A) and fission track dating (B) in the estimation of denudation and uplift rates relative to the landsurface. In (A) the age of mineral at the surface is determined by the time elapsed since it passed through the critical isotherm at which its radiometric 'clock' was set. This isotherm, and therefore depth, varies for different minerals. The rate of denudation is calculated by dividing depth (D) by age. In (B) the estimation of denudation using fission track dating of apatite is illustrated. If we assume that the geothermal gradient is 25°C km⁻¹ (as indicated) and the age of apatites exposed at the surface is 10 Ma then the mean denudation rate is 3400 (m)/10 = 340 m Ma⁻¹ (source: Summerfield, 1990, fig. 15.4).

the sites of former channels of the Lachlan and its tributaries.

2. The post-Oligocene rate of denudation was estimated by calculating volumes of sediment wedges that had been derived from catchments of a known area.
3. Rates of late Cretaceous and Permian denudation were derived either from sediment volumes, or from fission track ages from granitoid rocks.

A relatively new technique that may have considerable potential for determining rates of general denudation over long time-spans is cosmogenic exposure dating (Dorn and Phillips, 1991). A number of *in situ* cosmogenic radionuclides and stable nuclides have been measured in natural exposed rock surfaces with a view to study both their *in situ* production and their use to determine rock erosion rates. Two stable nuclides (^{3}He and ^{21}Ne) and five radionuclides (^{10}Be, ^{26}Al, ^{36}Cl, ^{14}C and ^{39}Av) are measureable in many rock types

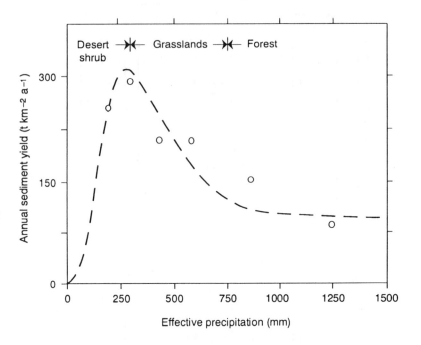

Figure 3.2
Variation of sediment
yield with climate as
based on data from small
catchments in the United
States (source: Langbein
and Schumm, 1958). The
data points represent the
means of considerable
data scatter so that their
validity is open to
question.

(see, for example, Lal, 1991; and Nishiizumi et al., 1993). These
nuclides build up in rocks due to the interactions of cosmic rays
with atoms in minerals by high-energy spallation, neutron-capture
reactions and muon-induced nuclear disintegrations. The rate of
accumulation depends on such measureable factors as altitude,
geomagnetic latitude, geometry of exposure to cosmic rays, and
also on time and on cosmic ray flux.

Global patterns of sediment yield: the climatic impact

Over the last four decades a range of attempts have been made to
assess rates of global denudation and the way in which rates vary
according to major environmental controls. Particular attention
has been paid to the rates at which rivers denude the landscape in
major climatic zones.

A highly influential study that related sediment yields to climatic
influences was that of Langbein and Schumm (1958). They used
American gauging station data relating to 94 catchments (mean
area 3885 km²) and reservoir sedimentation data from 163 catchments (mean area 78 km²). They found that sediment yields reached
a peak of 198 mm 1000 a^{-1} at the semi-arid/grassland precipitation boundary (mean annual precipitation of c.300 mm at a mean
annual temperature of 10°C). Sediment yield minima occurred in
very dry regions and in more humid ones. The explanation given
for this pattern (shown in figure 3.2) is that at effective precipitation

Plate 3.1
Although in hyper-arid areas rates of fluvial erosion are probably low because of the very limited availability of water to produce runoff, even in an area like Death Valley, California, USA, where the rainfall is less than 70 mm per year, fluvial erosion can be significant on erodible substrates.

levels of less than about 300 mm a^{-1} there appears to be insufficient runoff to produce maximum erosion (though see plate 3.1), whereas above it the erosive effects of increased runoff are more than counteracted by the prescence of a more or less continuous vegetation cover. The precise position of the peak varies with mean annual temperature, because rainfall effectiveness becomes less as temperatures rise.

Some more recent studies have tended to confirm the general trend of the Langbein and Schumm curve. For example, Inbar (1992) has analysed sediment yield data for four areas with a Mediterranean climatic regime: Israel, Spain, California and Chile. He infers from these data that there is a decrease of sediment yield with increasing precipitation, with a sediment yield that peaks with 300 mm of annual rainfall in the case of Israel and Spain and with 400 mm in the case of Chile and California (figure 3.3).

On the other hand, Yair and Enzel (1987), working in the Negev, were critical of the Langbein and Schumm curve as it relates to the very driest conditions (p. 133):

The prevailing idea that the denudation rate increases from zero to about 300 mm average annual rainfall is based on the assumption that surface properties are on the whole uniform within this wide range of climatic conditions and that erosivity is mainly controlled by the amount of annual rainfall. This assumption encounters serious difficulties when applied to the northern Negev, where a converse relationship seems to fit the reality better. . . . The detailed study of the present-day processes clearly indicates that surface properties, namely, the ratio of bare bedrock outcrop to soil cover, play a predominant role in runoff generation and in erosion processes. Runoff and erosion are positively related to this ratio. Since it is higher in arid than in adjoining semi-arid areas, sediment yields are higher in the former than in the latter areas.

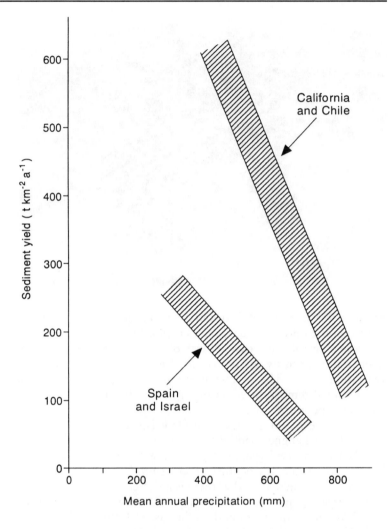

Figure 3.3
The decreasing rate of sediment yield with increasing precipitation in Spain–Israel and California–Chile (source: Inbar, 1992, fig. 6).

A further attempt to relate the global pattern of denudation to climatic controls was made by Fournier (1960). He used data for 96 streams (the majority of which were from the USA) and related their rates of denudation to the amount and seasonality of their precipitation. He then extrapolated from these data (which included no catchments from either South America, or Africa or Australia) to obtain a global pattern of denudation (figure 3.4). He wrote as follows about the zones where maximum denudation was to be expected (p. 185) (my translation):

The maxima of erosion are located in places where there is a combination of a high seasonality of rainfall with important amounts of precipitation. They are also located in mountainous regions. Calculated erosion exceeds 1000 t km^{-2} a^{-1} under a tropical or subtropical climate with a well-marked dry season (e.g. Africa, S.E. Asia, the extreme north of Australia and the easternmost part of South America). Equally this value is exceeded in

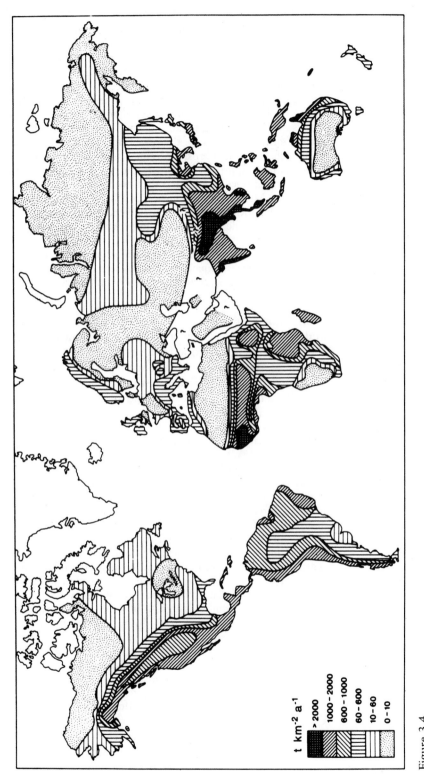

Figure 3.4
Global rates of denudation in t km⁻² a⁻¹ (source: Fournier, 1960, fig. xv, with modifications).

mountains where the climate is aggressive: the Rocky Mountains (semi-arid climate or excessively wet climate on the western slopes of the coastal ranges); the southern part of the Andes, the mountains and plateaux of Mexico, the mountains of West Africa, the mountains of Madagascar, the mountains of Annam, Himalaya and Szechwan (all these mountains are situated in the tropical or subtropical zones); those mountains of Asia with a semi-arid climate. Under a continuously wet tropical climate, erosion is much less strong because the apportionment of the rains is more regular during the course of the year. Under a tropical climate with a dry season, the calculated erosion declines as the abundance of rain diminishes. With the approach of a desert, the entrainment of earth over a long distance diminishes rapidly and then becomes nil.

Under a Mediterranean climate, a subtropical climate with summer drought, climatic aggressivity again engenders a high rate of erosion. It is the same under the subtropical climate with weak winter rainfall in the Far East.

Finally, it is above all in the temperate zone and in the cold zone that there are the biggest expanses of land where the entrainment of soil over a very long distance becomes slight, mountainous zones making the only exception.

Strakhov (1967, p. 13), on the basis of the analysis of discharge and sediment data from a mere 60 catchments, indicated that there was a great variation in the average intensities of mechanical erosion (c.500 times), with a range from 4 to 2000 t km^{-2} a^{-1}. He also used his limited data set to produce a map of the global pattern of erosion (figure 3.5) and identified a highly distinctive geographical pattern:

Two broad parallel zones with fundamentally different indices are recognised. The first embraces the temperate moist belt in the northern hemisphere, with 150–600 mm annual precipitation in North America, Europe, and Asia. Its southern boundary is the annual +10°C isotherm. The general intensity of mechanical denudation here is small: most commonly 10 tons/km^2, rarely 10–15 tons/km^2, and in only a single part of North America at a level of 50–100 tons/km^2. The second zone embraces parts of North and South America, Africa, and south-eastern Asia, adjoining the subtropical and tropical moist belt. This corresponds almost exactly to the region between the +10°C isotherm in the northern hemisphere and the same isotherm in the southern hemisphere; over a great part of this region the average annual temperature does not fall below 20°C, and it generally holds at 22–23°C; rainfall is high, 1200–3000 mm per year. The intensity of mechanical denudation is markedly greater than in the North Temperate Zone; most commonly it holds at a level of 50–100 tons/km^2, but in a number of places it rises to 100–240 tons/km^2, and in south-eastern Asia it averages 390 tons/km^2, and in the basins of the Indus, Ganges, and Brahmaputra, the value is even greater at 1000 tons/km^2 and more.

This pattern is, therefore, very different from that envisaged by Corbel (1964), who postulated that cold mountainous regions and glaciated regions had the highest rates of denudation.

One further study of global patterns of sediment yield undertaken in the 1960s was that of Holeman (1968). He calculated

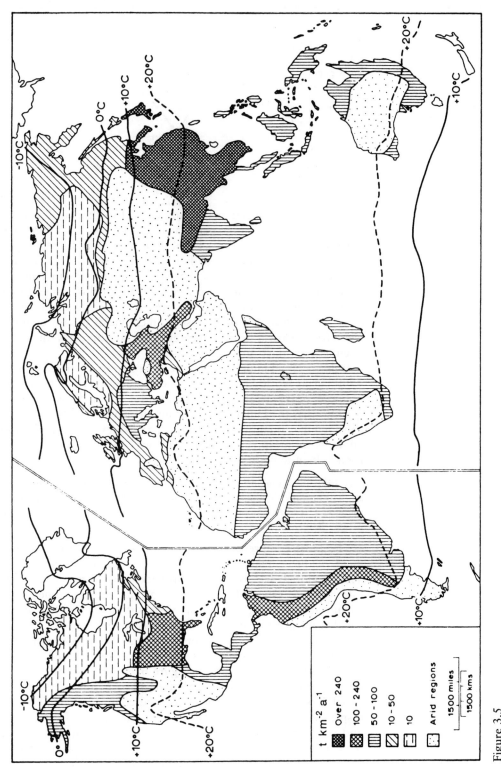

Figure 3.5
Global rates of mechanical denudation in t per km^{-2} a^{-1} per year (source: Strakhov, 1967, fig. 5, with modifications).

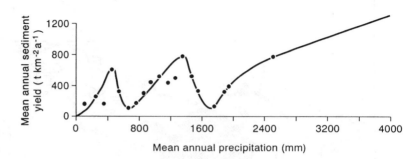

Figure 3.6
*Relationship between
mean annual sediment
yield and mean annual
precipitation (source:
Walling and Kleo, 1979).*

that the amount of sediment being carried to the world's oceans
was only 20.2×10^9 tons per year, compared with a figure of 64
$\times 10^9$ proposed by Fournier (1960) – a three-fold discrepancy. In
common with Strakhov and Fournier, and some later workers,
however, he appreciated the importance of the Asian landmass
which contributes some 80 per cent of the world sediment input
to the oceans from about one-quarter of the world's land surface
area. He believed that the lowest sediment yields for any continent
were from Africa, which generated on average less than one-
twentieth as much material as Asia.

Wilson (1973) undertook a more extensive study, based on some
1500 basins worldwide, and produced a new curve with two pronounced peaks at about 760 mm and 1768 mm. He suggested
(p. 339) that the Langbein and Schumm curve could 'no longer be
considered as quantitatively correct', and threw some doubt upon
attempts to interpret sediment yield rates solely in terms of some
broad climatic control (p. 339):

It is difficult to imagine any single variable that would explain more than
a small amount of the variation in sediment yield observed throughout
the world. If one such variable exists, it is probably a function of land use
rather than of climate.

The most comprehensive attempt of the 1970s to relate sediment yield to climate was that of Walling and Kleo (1979). Using
a large global data base they have identified three zones where
rates may be especially high: the seasonal climatic zones of the
Mediterranean type, monsoonal areas with large amounts of seasonal tropical rain, and semi-arid areas (figure 3.6).

A more recent attempt to produce a global map of suspended
sediment yield is that made by Walling (1987), based on more
than 1500 measuring stations (figure 3.7). The pattern relates to
the sediment yields from intermediate-sized basins of around 10^4
to 10^5 km², and is both generalized and complex. The high yields
displayed for the Mediterranean area, south-west United States
and parts of East Africa may possibly be related to the presence
of semi-arid climatic conditions and the precipitation control suggested by the Langbein and Schumm curve. On the other hand,

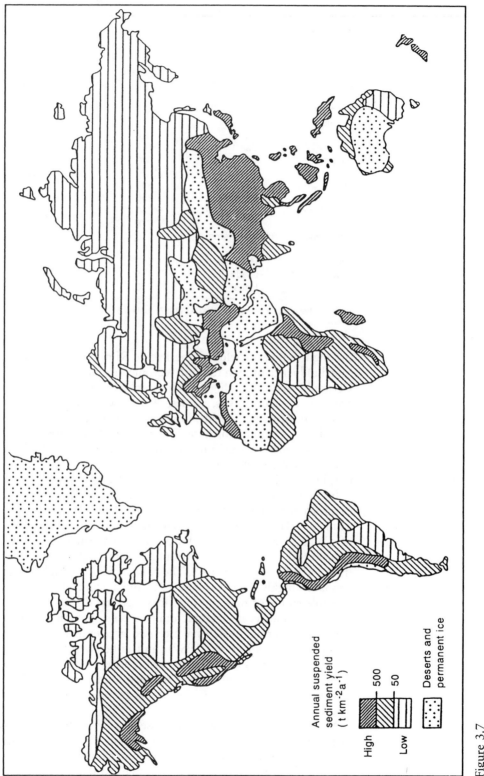

Figure 3.7
A global map of the pattern of specific suspended sediment yield based on data from over 1500 measuring stations (source: Walling, 1987, fig. 4.4).

Annual suspended
sediment yield
(t km^{-2}a^{-1})

High

500

50

Low

Deserts and
permanent ice

Table 3.2 Maximum values of mean annual specific suspended sediments yield for world rivers

Country	River	Drainage area (km²)	Mean annual suspended sediment yield (t km⁻² a⁻¹)
People's Republic of China	Huangfuchuan	3199	53,500
	Dali	96	25,600
	Dali	187	21,700
Taiwan	Tsengwen	1000	28,000
Kenya	Perkerra	1310	19,520
Java	Cilutung	600	12,000
	Cikeruh	250	11,200
New Guinea	Aure	4360	11,126
North Island,	Waiapu	1378	19,970
New Zealand	Waingaromia	175	17,340
	Hikuwai	307	13,890
South Island,	Hokitika	352	17,070
New Zealand	Cleddau	155	13,300
	Haast	1020	12,736

Source: after various sources in Walling (1987) table 4.5.

the high sediment yields of the Pacific Rim may reflect the combined influence of high rainfall, tectonic instability and high relief. Certainly, relief is a highly important control, with high steep areas in the Himalayas, Andes, Alaska and the Mediterranean lands producing high yields. Low rates are evident for much of the old shield areas of northern Eurasia and North America, with their low relief and resistant substrates, and for equatorial Africa and South America, which reflect the presence of subdued topography and the dense cover of rainforest vegetation.

In some parts of the world, suspended sediment yields can be extraordinarily high (table 3.2) (Walling, 1987), exceeding 10,000 t km⁻² a⁻¹. The highest value comes from the Huangfuchuan River of China, a 3199 km² catchment with a mean annual yield of 53,000 t km⁻² a⁻¹. This is a tributary of the Yellow River, and it drains a gully region of loess. The reasons for such high values are probably highly varied (plate 3.2), and include the presence of erodible stores of material (e.g. loess, glacial drift, volcanic ash), high relief and recent tectonic uplift, intense human pressures on the ground surface which have caused accelerated late Holocene erosion (Milliman et al., 1987), and either a semi-arid climate (with a limited vegetation cover) or highly erosive rainfall regimes dominated by tropical storms or very high annual totals. It is also worth noting that all the basins in table 3.2 are relatively small.

Most of the data we have discussed so far in this section have primarily involved suspended sediment data. Comparative data on bed-load transport fluxes under different climatic conditions are remarkably sparse. Not least this applies to data for the world's

Plate 3.2
Some of the fastest rates of denudation occur in the high mountains of South Asia. The Hunza River in the Karakoram of Pakistan has a load of around 5000 tonnes km^{-2} a^{-1}. This can be attributed to such factors as the high available relief, stores of Pleistocene debris, glaciers in the catchment, and severe frost and salt weathering.

arid zones where, because of infrequent and unpredictable floods, there is a paucity of reliable field data. A recent study by Laronne and Reid (1993) in the Negev Desert, Israel, based on monitoring of bed-load movement in flash flood events, indicates just how crucial continuous bed-load monitoring in such environments could be. Their data show that ephemeral desert rivers could be as much as 400 times more efficient at transporting coarse material than their perennial counterparts in humid zones. They argue that this may be because of the relative lack of bed armouring in desert streams. They suggest that the poor or non-existent development of armoured layers could be a function of the rapid rise and fall in water discharge that is characteristic of flash floods, the high rates of sediment transport and the extended intervals of dormancy between flood events. These characteristics reduce the tendency towards the size-selective transport of clasts and the winnowing of fines that are each thought to promote armouring in perennial channels.

Some regional patterns of denudation

Milliman (1990, 1991) has drawn attention to the particularly high amounts of sediment transported to the oceans from southeastern Asia and Oceania (figure 3.8). These areas account for only about 15 per cent of the land area draining into the oceans, but contribute about 70 per cent of the flux of suspended sediment. In contrast, rivers draining the Eurasian Arctic, a basin area similar in size to southern Asia and Oceania combined, contribute about two orders of magnitude less sediment.

Milliman and Syvitski (1992) believe that the sediment fluxes from small mountainous rivers, many of which discharge directly onto active continental margins, may well have been largely neglected in the past, but that their contribution of sediment to the oceans may be remarkably important. They point out, for example, that in North America the loads of the Sustitna, Cooper, Stekine and Yukon collectively exceed that of the Mississippi. In Europe, the Semani River of Albania discharges more than twice as much sediment as do the Garonne, Loire, Seine, Rhine, Weser, Elbe, Oder and Vistula combined.

A particularly large proportion of the sediment yield to the oceans comes from the world's highest mountain chain – the Himalayas. Estimates of the annual sediment yield from the continents to the oceans for the world as a whole range from 2.7 to 4.6 km^3 a^{-1} (Hales, 1992). More than 1.1 km^3 a^{-1} comes from the collision-created Himalayas, and is deposited in the back arc basins of east Asia or in the great Bengal Fan.

By contrast, one continent where overall rates of denudation appear to be particularly slow is Australia (figure 3.9). This is significant as it is also a continent where landscapes appear to be particularly old and persistent. Australian researchers have provided evidence that a significant component of the landscape of this continent has its origin beyond the start of the Quaternary and, often, before the start of the Cenozoic. Although some of the ancient landscapes may be exhumed features (e.g. the Devonian reef of the Napier and Oscar Ranges in North West Australia has been exhumed from beneath a later Permian (?) sedimentary cover (Goudie et al., 1990)), many landscapes have remained as subaerial features since their inception tens or hundreds of millions of years

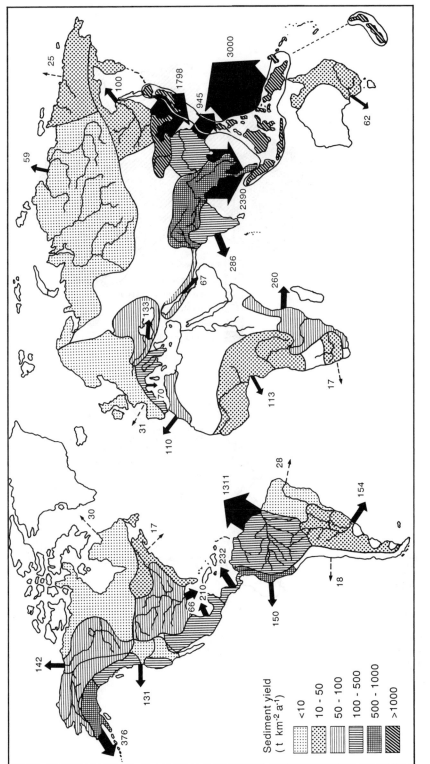

Figure 3.8
Annual fluvial sediment flux from large drainage basins to the oceans. Numbers in millions of tons per year with size of arrows proportional to the numbers (source: Milliman, 1990, fig. 4).

Sediment yield
(t km⁻² a⁻¹)

<10
10 - 50
50 - 100
100 - 500
500 - 1000
>1000

(a)

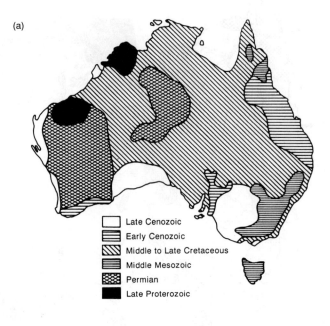

Late Cenozoic
Early Cenozoic
Middle to Late Cretaceous
Middle Mesozoic
Permian
Late Proterozoic

(b)

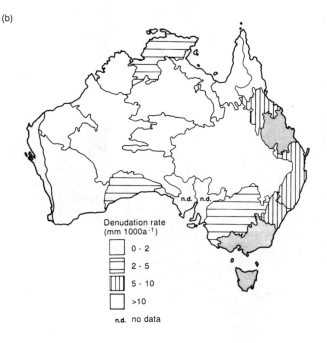

Denudation rate
(mm 1000a⁻¹)

0 - 2
2 - 5
5 - 10
>10

n.d.   no data

Figure 3.9
*Long-term landscape
evolution in Australia
(source: Gale, 1992, fig.
12):*
*(a) The age of inception
of drainage.*
*(b) Cenozoic denudation
rates in relation to
major physiographic
divisions in mm 1000
a⁻¹.*

ago, and may even date back to the Proterozoic. Most estimates of Mesozoic and Cenozoic erosion rates are less than 10 m per million years, and that for the Yilgarn Block, at a few tens of centimetres per million years since the Miocene, must be one of the lowest denudation rates for any land surface in the world. Gale (1992) puts forward some possible reasons for the low rates of denudation in Australia: large parts of the country have been orogenically relatively stable since the Permo-Triassic, producing low relief and limited potential energy; the country is relatively arid and has areas of internal drainage; and the country suffered from only very limited glaciation in the Pleistocene.

The low rates of denudation and drainage incision even apply to some of the highest portions of south-eastern Australia. Thus, Young and McDougall (1993) have compared the nature of drainage systems and channel gradients beneath dated basaltic lavas with those present in the contemporary landscape, and suggest that plateau surfaces in southern New South Wales have in general only been lowered by 2 to 5 mm 1000 a^{-1}, and that most major streams have only incised since the Miocene by 5–18 mm 1000 a^{-1}.

In terms of both rates of denudation and physiographic setting, New Zealand forms a striking contrast to Australia. Whereas Australia is for the most part tectonically quiescent and dominated by old, low relief land surfaces, New Zealand is a highly dynamic tectonic environment with major mountain chains and high inputs of precipitation. It also experiences some of the highest fluvial denudation rates in the world with quite large catchments in the Southern Alps having erosion rates that may be in excess of 15,000 t km^{-2} a^{-1} (Adams, 1980; Griffiths, 1979). These rates are all the more remarkable when one remembers that anthropogenic pressures on many of the highest mountainous catchments have been slight. High available relief, intense frost and nivation processes, glacierization, rapid rates of uplift, susceptible rocks, and a large store of Pleistocene debris, may all contribute to these high rates, but the prime control appears to be the very large quantities of precipitation that fall in the high mountains (often 5000 or more mm per annum). Indeed, in his analysis of precipitation and sediment yield data, Griffiths (1979) found a simple positive power law relationship between suspended sediment yield and mean annual rainfall.

For the United States, Judson and Ritter (1964) used data on both dissolved and suspended load to calculate rates of regional denudation. Their estimates for major hydrological regions were as follows:

	mm 1000 a^{-1}
Colorado	165
Pacific slopes (California)	91
Western Gulf	53

Table 3.3 Erosion rates in upland areas of the south-western Basin and Range Province of the USA inferred from comparison of active and relict basalt-capped erosion surfaces

Location	Minimum age of relict erosion surface	Maximum average downwasting rate (mm 1000 a^{-1})
Buckboard Mesa (S. Great Basin)	2.82	47
Cima Volcanic Field (Mojave Desert)	4.48	11
	3.88	28
	3.64	25
	0.85	28
	0.67	30
Lunar Crater Volcanic Field (Central Great Basin)	2.86	8
	1.08	11
Reveille Range (Central Great Basin)	5.70	31
	3.79	16
White Mts. E. Flank (SW Great Basin)	10.8	24
	10.8	20

Source: modified after Dohrenwend (1987) table 4, p. 331.

	mm 1000 a^{-1}
Mississippi	51
S. Atlantic and E. Gulf	41
North Atlantic	48
Columbia	38
Mean	61

They were impressed by what this might mean in terms of long-term denudation and geomorphological evolution (p. 340):

Taking the average height of the United States above sea level as 2300 feet and assuming that the rates of erosion reported here are representative, we find that it would take 11 to 12 million years to move to the oceans a volume equivalent to that of the United States lying above sea level. At this rate there has been enough time since the Cretaceous to destroy such a landmass six times.

An interesting method of obtaining long-term denudation rates for a particular part of the USA was employed by Dohrenwend (1987). Working in the Basin and Range Province he argued that the region's complex history of late Cenozoic tectonic activity and climatic change precluded the use of modern sediment yields as reliable indicators of long-term erosion rates. He suggested instead that a more reliable approach would be to calculate average downwasting rates by measuring the average vertical distance between a basaltic lava-flow capped erosion surface and the level of an adjacent modern erosion surface, and dividing this height difference by the K-Ar date of the lava flow. The estimates for the Great Basin and Mojave Desert, covering periods from 0.67 to 10.8 million years, range between 8 and 47 mm 1000 a^{-1} (table 3.3).

A recent evaluation of rates of fluvial denudation in Canada has been made by Ashmore (1993). With the exception of the Western Cordillera, where rates may be as high as 300 t km^{-2} a^{-1}, it is unusual to find annual specific yields exceeding 100 t km^{-2} a^{-1}. Yields from streams with prairie sources in the eastern prairies are typically low and often much less than 10 t km^{-2} a^{-1}, with the notable exception of the small steep basins incised into the erodible shale of the Manitoba Escarpment, where specific yields are comparable to those of many Cordilleran basins. Ashmore makes some important statements about the relationship between basin area and sediment yield in Canada, which may have wider significance (p. 199):

Interpretations of regional sediment load data for various regions of Canada suggest that conventional models of sediment routing seldom apply. The land surface is often either not well integrated with the main stream system or experiences extremely low erosion rates. Streams are frequently incised into glacial deposits or incompetent bedrock. As a result, streambank and valley side erosion are often the dominant sediment sources. This produces a positive relation between specific sediment yield and basin area over several orders of magnitude of drainage area, which is counter to the conventional pattern. This outcome, largely a legacy of Quaternary glaciation, is likely to persist and relaxation times are probably of the order of 10^5 years.

Sediment loads of British Rivers

The suspended loads of British rivers are highly variable, with a spread from less than 1.0 t km^{-2} a^{-1} to almost 500 t km^{-2} a^{-1}. Walling and Webb (1981) regard a value of 50 t km^{-2} a^{-1} as typical of British rivers, and regard this as being low by global standards. This is probably because of two particular characteristics of the British environment: the low peak intensities of rainfall, and the relatively dense vegetation or sod cover (Moore and Newson, 1986). Estimates of long-term sediment yields based on reservoir surveys are few in number for the UK but seem to give figures of the same order of magnitude as the river sediment yield data (probably around 30 t km^{-2} a^{-1}).

The available data do, however, show a considerable range, and Labadz et al. (1991) show values with a hundred-fold span, from 2.1 t km^{-2} a^{-1} to 205.4 t km^{-2} a^{-1}. Part of the variation is explained by the fact that different workers have used different means of calculating rates of sediment yield from the reservoir sedimentation data, employing, for example, different density and shrinkage values, and with some ignoring the organic fraction. This latter tendency is imprudent given that in some parts of highland Britain the amount of material derived from peat erosion is substantial.

The rates of sedimentation in some British reservoirs are appreciable and cause a loss of capacity. Data on this problem as it relates to reservoirs in the southern Pennines are provided by Butcher et al. (1993), who present values for capacity loss percentages per century that range from 0.3 to 47.1. Loss of capacity is especially large for those reservoirs that receive a large input of low density organic material derived from the erosion of peat. The organic contents for reservoir sediments in the area range from 7.4 to 63.1 per cent.

Walling and Webb (1981, p. 167), recognizing the relative sparsity of data, allow themselves a few tentative generalizations about the pattern of suspended sediment yield in Britain:

Loads in excess of 100 t km^{-2} a^{-1} would seem to be associated primarily with upland areas receiving annual precipitation greater than 1000 mm and, within these areas, with small- and intermediate-sized catchments where sediment delivery ratios will be relatively high. Conversely low suspended sediment yields (< 25 t km^{-2} a^{-1}) would appear to reflect low annual precipitation (e.g. River Welland and River Nene), large basins where sediment delivery ratios will be relatively low . . . and low relief. . . . Very low suspended sediment yields (< 5 km^{-2} a^{-1}), represented by the East Twin catchment on the Mendip Hills and the Ebyr N. and Ebyr S. catchments in central Wales, may be accounted for in terms of the small headwater areas involved, the resistant bedrock and the essentially undisturbed conditions found in these upland areas.

Data on bed-load transport by British rivers are even sparser. However, Newson (1981, p. 77) suggests that 'the high residual proportion of coarse materials in British upland channels means that the majority of sediment transport occurs as bedload', whereas in non-upland areas the proportions are reversed. Certainly, in mid-Wales bed load does appear to be more significant than suspended load (Reynolds, 1986), but in the Pennines, another upland, bed load appears to be of very limited importance (Carling, 1983).

Relief and denudation rates

The vigour of denudation in mountains has long been appreciated. Reclus (1881, p. 53), for example, pointed out that in spite of their imposing bulk, they were in a sense mere Goliaths in the face of the erosional David:

And yet these enormous masses, mountains piled upon mountains, have passed away like clouds swept along the sky by the wind; the strata four or five thousand yards thick, which the geological section of rocks shows us had formerly existed, have disappeared to enter into the circuit of a new creation. It is true that the mountain still appears formidable to us, and we contemplate with admiration mingled with alarm the superb peaks penetrating far away beyond the clouds, into the icy atmosphere of space. So lofty are these snow-clad pyramids that they conceal one half of the sky from us; from below their precipices, which in vain our glance tries to grasp, make us dizzy. Yet all this is but a ruin, mere debris.

That available relief should be a major control of the speed at which the landscape is worn away is self evident. As Ahnert (1970, p. 243) wrote:

The influence of available relief upon the rate of denudation is a fact familiar even to the casual observer who compares the rushing power of a mountain torrent with the quiet flow of a lowland brook. The greater the relief, the steeper are, as a rule, the slopes, and the more rapid is the downhill transport of waste and thus the downwearing, or denudation, of the land. Equally obvious is the fact that uplift tends to increase the relief and thus exerts an indirect influence upon the rate of denudation, too.

Hewitt (1972) has maintained (p. 18) that 'the vigour of erosion, and frequency of larger erosional events, is never so extensive and uniformly productive of high erosion rates as in mountains'. However, he also pointed to the causes of such vigour and stressed that available relief is more important than altitude *per se* (p. 19):

The sources of the high energy condition have generally been identified as 'relief energy' and extremes of climate. The positive mass of the solid earth above sea level represents gravitational potential energy derived from those tectonic processes which govern the broad configuration and development of the crust. Moisture deposited at higher altitudes has greater potential energy. Yet positive mass and altitude are not themselves sufficient to produce vigorous geomorphic activity. Geomorphic conditions on the high plateau and interior basins of Central Asia, the North American Cordillera or the Altiplano of Andean South America are not equivalent to those of the true mountain areas which border them. More important than altitude is available relief. Either differential uplift must create steep mountain fronts or, as is more usual, the expenditure of positive mass must have proceeded to a point where dissection of the mountain block is fairly complete. The availability of potential energy for geomorphic processes depends most directly upon the local relief and distribution of slope angles. For a given range of altitude it is the more thoroughly dissected mountains which exhibit the higher erosion rates. If we compare the hypsometry of the Pamirs and the Karakoram Himalaya the former are found to have greater positive mass. The latter is more deeply and widely dissected although having a greater array of high peaks and crestlines. The two are broadly similar in glacierisation, total relief and climates. It seems, therefore, that degree of dissection explains the apparently much higher denudation rates for the Karakoram.

Anhert himself sought to quantify the relationship between relief and rate of denudation by selecting a range of mid-latitude basins (from North America and Europe) which were from several hundred km^2 to over 100,000 km^2 in area, and occurred in locations where the mean annual precipitation ranged from less than 250 mm to over 2500 mm. He obtained data on both the solid and dissolved components of the load to establish an overall rate of denudation. He found a statistically highly significant correlation coefficient of 0.98 between mean relief and mean denudation rate, and suggested that relief was a far more important control

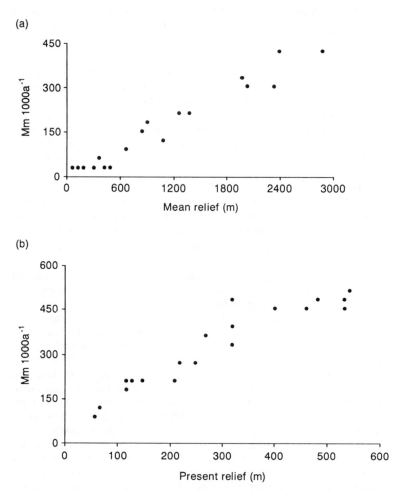

Figure 3.10
*The relationship between
relief and denudation rate
in mm 1000 a⁻¹:*
*(a) In mid-latitude
    medium-sized drainage
    basins (source:
    modified from Ahnert,
    1970).*
*(b) For Hydrographer's
    Volcano, Papua
    (source: modified from
    data in Ruxton and
    McDougall, 1967).*

of denudation rate than precipitation. The linear relationship he
established is shown in figure 3.10.

Corbel (1964, p. 399) was among those authorities who stressed
the major differences that exist between rates of erosion in mountainous areas and those in plains. He provided the following rates:

Plains	*mm 1000 a⁻¹*
Cold climate	18
Temperate climate	46
Hot and dry climate	11
Hot and moist climate	21
MEAN	22
Mountains	
Cold climate	385
Temperate climate	110
Hot and dry climate	228
Hot and moist climate	92
MEAN	206

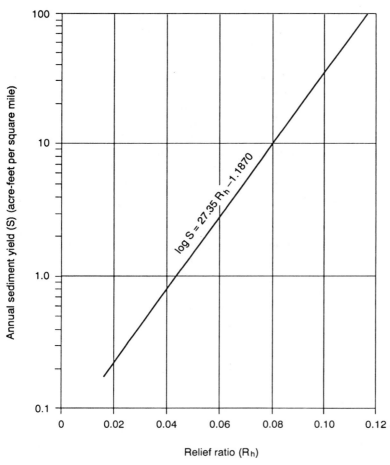

Figure 3.11
The relationship between annual sediment yield per unit area and relief ratio (i.e. maximum basin relief divided by length) in respect of drainage basins of approximately 1 sq. mile in area underlain by sandstone or shale in the semi-arid western United States (source: Schumm, 1963, fig. 1).

The above figures comprise both mechanical and chemical denudation.

Other authors have found a broadly similar tendency. Schumm (1963) identified an exponential increase of annual sediment yields with relief ratio for basins of about 1 square mile (2.6 km²) developed on sandstone and shale in the semi-arid south-western United States (figure 3.11). Ruxton and McDougall (1967), using data on the dissection of a dated andesitic stratovolcano from Papua, found a linear relationship between the average denudation rate and available relief (figure 3.10b).

Phillips (1990) provides an interesting analysis of the relative importance of different controlling factors at the global scale, and concludes by stressing the very large role played by the relief factor in explaining the variability in rates of erosion in different locations. Likewise, Pinet and Souriau (1988), after an analysis of large drainage basins, found that of all the factors that they considered in a multiple correlation analysis, relief (mean basin elevation) was the most important control of mechanical denudation of the continents and that precipitation was the main control of rates of chemical denudation.

It is however, important to stress that it is probably relative relief rather than elevation (altitude) per se which is the crucial control of rates of denudation. There need be no relationship between elevation and either mean slope gradient or local relief. This is a point that has been made in the context of southern African denudation rates by Summerfield (1991b).

These observations of the relationship between relief and rates of denudation can, as Schumm (1977, p. 21) pointed out, be viewed in another manner. They imply that sediment yields will also change through time, as high relief can be seen as analogous to the geomorphological stage of 'youth'. As the landscape is progressively worn down, the reduction in relief will mean that there will be a reduction in the quantity and size of the sediment load derived from the drainage system. This is a view that has been developed further by Fraser and de Calles (1992) who examined the relationship between the progress of drainage network development through time and the rates of erosion and sediment yield. Their view of drainage network development was as follows (p. 234):

Development begins with a period of *initiation* during which a shallow, skeletal drainage network is established. Elevational differences between source and basin core are at a maximum at this time, but relief in the drainage basin itself is at a minimum. Headward extension of the trunk stream and its major tributaries occurs during the period of *elongation*, and with the onset of *elaboration*, minor tributaries are established in previously undissected areas. *Maximum extension* is reached when the whole of the drainage basin is occupied by stream courses that are separated by narrow interfluves. Progressive reduction of relief sets in and a gradual loss of tributaries occurs during the period of *abstraction* until only minor drainage courses remain.

However, the rate of drainage network evolution gradually slows down (unless interrupted by tectonic, climatic or other changes) and so does the rate and size of sediment yield (p. 235):

The rate of drainage network evolution slows with time. A critical shear stress is required to initiate erosion of channels. This shear stress is controlled by discharge and water depth, which are dependent, under conditions of constant rates of precipitation, on the size of the catchment basin of the channel, and by channel slope. As stream channels extend into the available area, they expand their catchments, but extension also triggers the development of higher order tributaries which act to reduce the catchment area available for each channel. Because the rate of erosion is proportional to the excess of the available contributing area over the minimum required to produce critical shear stress, the rate of headward erosion will decline as the drainage network expands and higher order tributaries are added.

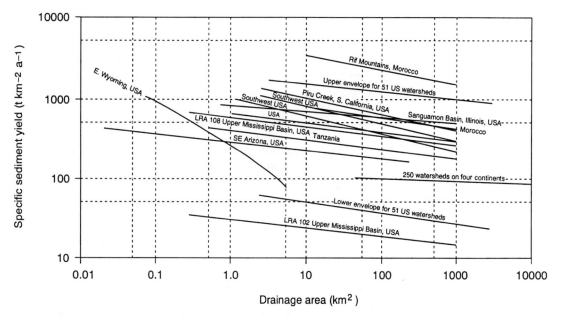

Figure 3.12
The conventional model of specific sediment yield in relation to drainage basin area (source: based on Owens and Slaymaker, 1992, fig. 1).

This development is crucial to the way in which sediment is liberated and stored in a catchment, and determines, *inter alia*, the rate and style of alluvial fan development (p. 236):

...we would normally expect that the supply of coarse sediment would be at a maximum when relief in the basin was greatest and local slopes were steepest. Under these conditions the gravitational component aiding downslope movement is greatest, overland flow into stream channels is most efficient, storm flow hydrographs peak faster, and higher, and stream competency is at a maximum. Relief and local slope steepness decline as drainage basins evolve, however, leading to a decrease in the size of clasts supplied to channels and a decrease in the carrying capacity of the streams. The gravitational component of downslope movement of sediment decreases so that the residence time of sediment on slopes increases.

Basin area, basin storage and sediment yield

It has frequently been observed that sediment yield is scale dependent. Yields tend to be greater per unit area for small basins than for large, at least up to basins 2000 km² in area (figure 3.12). The basin area effect may largely disappear when very large basins are considered because the topographic controls tend to become less variable between basins at this scale. Schumm (1963) has proposed the following relationship for basins in the south-west United States:

$$S \propto A^{-0.15}$$

where S is the mean annual sediment yield (in acre-feet per square mile) and A is the basin area (in square miles). Chorley et al. (1984, p. 53) suggest that for otherwise similar small basins a halving of the area may increase sediment yield by ten times. They believe that the inverse relationship is due to the following properties of smaller basins in relation to larger ones:

1. They commonly have steeper stream and valley side gradients, which encourage more rapid rates of erosion.
2. Small streams have a lack of floodplains, which gives less opportunity for sediment storage within the catchment, resulting in a higher sediment delivery ratio.
3. Small basins are more prone to being blanketed by one high intensity storm event, giving high maximum erosion rates per unit area.

However, the inverse relationship between sediment yield and basin area suggested above is by no means invariable. For example, Church and Slaymaker (1989) found a positive correlation between sediment yield and basin area in glaciated basins in British Columbia, Canada, at all spatial scales up to $3 \times 10^4 \, km^2$ (see also Owens and Slaymaker, 1992). They explained this result in terms of secondary remobilization of Quaternary sediments along river valleys. On the other hand, Reneau and Dietrich (1991), in their study of catchments in the Oregon Coast Range, USA, found no systematic relationship between sediment yield and drainage area for basins varying in size by three orders of magnitude. They suggest that a reason for these three different sets of results is that different environments will behave differently. For example, basins that have suffered from severe human disturbance may undergo severe headwater erosion and substantial aggradation lower down. Such basins would follow the relationship proposed by Schumm. Basins that have a large store of glacially derived sediment waiting to be evacuated, may have the sort of relationship proposed by Church and Slaymaker, whereas the occurrence of an approximate equilibrium between hill-slope erosion rates and rates of stream sediment movement may occur in areas where neither glaciation nor human activities have caused major fluctuations in sediment storage.

Erosion on a hillslope does not necessarily lead to the eroded material immediately entering the stream channel or to it being transported along the stream to its mouth. This concept has been well expressed by Phillips (1991, p. 231):

Many things can happen to a soil particle or a piece of clastic debris once it has been eroded by water from a hillslope. In general terms, it can be deposited further down the hillslope (sometimes only a short distance away), or transported to a stream channel. Once in the stream system it may be stored as alluvium in the channel or on the floodplain, or carried downstream and out of the drainage basin. Any and all of these things may happen to a single particle, perhaps more than once.

These simple ideas have long been known, but geomorphologists have recently become aware that sediment storage within a drainage basin may be the single most important aspect of fluvial sediment systems in terms of determining fluvial system response to environmental change. In recent years numerous authors have shown that: (1) large proportions of the sediment produced by erosion, mass wasting, and other processes within a basin are stored for significant amounts of time (≥ 1 yr) within the basin; (2) changes in storage may offset or overshadow changes in sediment production in influencing river sediment yield, and (3) storage may be more sensitive than yield to environmental changes.

Amongst other things, this means that it is unwise to use sediment yield data from gauged catchments as a means of estimating the amount of erosion going on *within* such catchments.

A classic area for comparing rates of hill-slope erosion and rates of fluvial sediment evacuation is the Piedmont region of the eastern USA. In the southern Piedmont, Trimble found that the mean sediment delivery ratio was only 6 per cent, with individual values for ten large basins ranging from 3.5 to 12 per cent. In a later study on the North Carolina Piedmont, Phillips (1991) found that sediment delivery ratios ranged from 7 to 16 per cent. He calculated that about three-quarters of the sediment produced by erosion on hill slopes is stored as colluvium.

The study of rates of sedimentation in lakes reveals a related phenomenon, namely that sediment yields are greatest in catchments which are small relative to the lake area (figure 3.13). Dearing and Foster (1993, p. 7) believe that this relationship is explained in part by the amount of sediment stored and by the number of erosional routes between slopes, stream channels and the lake. In relatively small catchments the storage component tends to be low while the number of erosional routes tends to be high. Their observations suggest that other things being equal, sediments from lakes with relatively small catchments (e.g. a catchment to lake ratio of less than 10) are more likely to be derived from slope or surface processes rather than from channel processes. Conversely, in relatively larger catchments (e.g. with a catchment to lake ratio of more than 10), fluvial processes become more important as the catchment becomes large enough to support a channel network.

Rates of denudation over millions of years

Because modern sediment yield data have a variety of problems which may render them unrepresentative of long-term trends, a variety of 'geological' methods has been developed to enable the calculation of denudation rates for periods of millions of years. Such methods include the use of fission track dating and the study

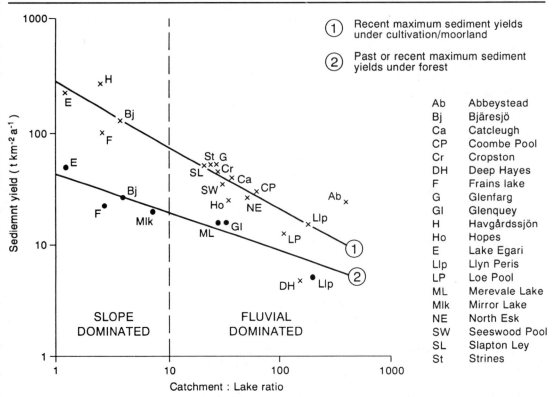

Figure 3.13   *Relationship between sediment yield and catchment to lake ratio in published studies (source: Dearing and Foster, 1993, fig. 2.2, with modifications).*

of sediment volumes in sedimentary basins. There are as yet insufficient studies of this type to permit one to make any very wide-ranging generalizations on this subject, but undoubtedly such studies have enormous potential geomorphological and geological significance.

It is perhaps instructive to compare rates of long-term denudation for two contrasting types of environment: on the one hand eastern North America, an area of ancient mountains and piedmonts with relatively modest relief, and on the other active zones of collision on plate margins (the European Alps and the Karakorams/ Himalayas), with very great relief. In eastern North America, Matthews (1975) calculated rates of denudation from sediment volumes offshore from a zone stretching between Georgia and Newfoundland. He suggested that there had been an average amount of denudation of some 2000 m since the start of the Mesozoic, implying a very slow rate indeed. Partial confirmation of this comes from a fission-track based technique applied to northern New England by Doherty and Lyons (1980). They also found a slow rate of about 1000 m 32 Ma^{-1} for the period from the Mesozoic to the Holocene.

   The contrast with the results from the young mountains formed
by relatively recent or current plate collision is stark. For the
European Alps, using Rb/Sr and K/Ar dates on biotites, Clark and
Jäger (1969) calculated a rate of denudation of 0.4 to 1.0 mm per
year (400–1000 mm 1000 a^{-1}), indicating that (p. 1159), 'a few
tens of kilometres have been removed by erosion during the Tertiary history of the Alps'. Even faster rates are indicated for the
Nanga Parbat area of northern Pakistan, where the fission track
technique suggests that over the last 1–2 million years the mean
regional erosion rate has averaged around 3–4 mm per year (3000–
4000 mm 1000 a^{-1}), whereas studies of basin sediment volumes in
the Cenozoic Foreland Basin of northern Pakistan indicate that
high rates of erosion (1–15 mm per year) have persisted over quite
large areas (60–2000 km^2) for quite long time periods (0.2–1.5
Ma^{-1}) (Burbank and Beck, 1991).
   The eastern North America data indicate rates of 10–30 mm
1000 a^{-1}, whereas those for the Alps and northern Pakistan indicate values of 1000–15,000 mm 1000 a^{-1}. These data indicate very
strongly the importance of tectonic situation for understanding
long-term rates of denudation.
   One way of trying to establish whether rates of erosion have
shown significant trends over periods of millions or tens of millions of years is to look at the rates at which sedimentation of
terrestrial material has taken place in the oceans. This can be
achieved by looking at the rate of accumulation of different terrestrial indicators (e.g. Al$_2$O$_3$). Davies et al. (1977) estimated
average rates of Cenozoic sedimentation for the Atlantic, Indian
and Pacific Oceans using cores from the Deep Sea Drilling Project
(DSDP), and found globally synchronous fluctuations. They found
that Palaeocene to early Eocene and late Eocene to early Miocene
rates were only a fraction of middle Eocene and middle Miocene
to Recent dates. They suggested that this pattern could be largely
explained by climatic change, and suggested that the lowest rates
of erosion were associated with periods of less runoff. Donnelly
(1982) found a six-fold increase in the rate of accumulation of
Al$_2$O$_3$ in Atlantic and Pacific sediments in the last 15 million years
and attributed this to a vastly increased denudation rate on the
continents. They argued that the comparability of pattern between
the Atlantic and Pacific largely ruled out a tectonic effect because
of the very different tectonic styles of the two oceans. They also
suggested that climatic change was probably responsible, and in-
voked the role of world-wide climatic deterioration, especially in
high and middle latitudes.
   However, Summerfield (1991a, p. 391) is very cautious about
drawing conclusions about the relationship between long-term and
present-day denudation rates:

Although we are living in a rather unusual period of geological time (the
Quaternary) characterized by rapid changes in eustatic sea level and major

oscillations in global climate, contemporary fluvial denudation rates, perhaps rather surprisingly, do not appear to be significantly different from the long-term average. The estimated present-day global mean of 43 m Ma^{-1} is somewhat higher than the long-term estimate of 11 m Ma^{-1} based on DSDP data, but the latter is very uncertain and is probably a significant underestimate of total continental denudation because it does not adequately incorporate continental margin sedimentation. Certainly, fission track estimates of mean denudation rates of around 30 m Ma^{-1} during the Cenozoic in regions of subdued relief are not incompatible with a global average in excess of 40 m Ma^{-1}. Moreover, the range of present-day denudation rates appears to be very similar to the range evident for the long term, with minima of around 1 m Ma^{-1} and maxima in excess of 5000 m Ma^{-1}. There is evidence that rates in the Late Cenozoic may have increased significantly in some regions as a result of active tectonism or climatic oscillations, and changes in areas of internal and external drainage must also have affected sediment input on to the continental margins.

Although at a global scale there is a broad correspondence between present-day and long-term rates of denudation, anthropogenic effects have clearly become overwhelming in some areas.

The rate of peneplanation

In the classic Davisian model, the cycle of normal erosion, a period of rapid uplift is followed by a long period of relative tectonic quiescence during which erosion gradually reduces the landscape to a low-lying, low relief plain – the peneplain. Schumm (1963) demonstrated that rapid pulses of uplift are the norm, and that modern rates of orogeny are much greater (c × 8) than the average maximum rate of denudation. He suggested that this gave some support to the Davisian notion of a period of rapid uplift, which allows little erosional modification of an area before the cessation of uplift. However, he also pointed out that rates of land surface reduction would be lessened by a combination of isostatic compensation for the amount of material removed by erosion, and by the fact that rates of denudation will decline as relief becomes more limited during the passage from youth through maturity to old age. Bearing such considerations in mind, he suggested that planation of about 1500 m of relief may require perhaps 15 to 110 million years.

Hales (1992) has attempted to calculate how long it would require, given the current hypsometric characteristics of the Earth and the current rates of sediment yield to the ocean, for the reduction of the continental crust to sea-level were orogeny and other uplifting processes not to occur. Using a figure for the mean altitude of continental crust above sea-level to be 623 m, he estimated that without any allowance for isostatic compensation the continental crust would be reduced to sea-level in about 32 million years. Allowing, as one needs to, for isostatic compensation the reduction to sea-level would take some 160 million years.

Plate 3.3
*Land-use change (bush
clearance) in Swaziland,
southern Africa, has
exposed the wedges of
colluvium that store
sediment along the
Mkhondvo River to severe
erosion in recent
centuries, producing great
dongas (badlands).*

Accelerated erosion associated with land-use change

Land-use changes executed by humans have made very substantial
changes to 'natural' rates of erosion (plate 3.3). Of prime importance has been the acceleration of rates caused by deforestation.
It is therefore prudent to take care with extrapolating modern
rates of erosion to give long-term 'geological' rates.

Forests protect the underlying soil from the direct effects of
rainfall, generating what is generally an environment in which
erosion rates tend to be low. The canopy plays an important role
by shortening the fall of rain drops, decreasing their velocity and
thus reducing kinetic energy. There are some examples of certain
types (e.g. beech) in certain environments (e.g. maritime temperate) creating large raindrops, but in general most canopies reduce
the erosive effects of rainfall. Possibly more important than the
canopy in reducing erosion rates in forest is the presence of humus
in forest soils (Trimble, 1988), for this both absorbs the impact of
raindrops and has an extremely high permeability. Thus forest
soils have high infiltration capacities. Another reason that forest
soils have an ability to transmit large quantities of water through
their fabrics is that they have many macropores produced by roots
and their rich soil fauna. Forest soils are also well aggregated,
making them resistant to both wetting and water drop impact.
This superior degree of aggregation is a result of the presence of

considerable organic material, which is an important cementing agent in the formation of large water-table aggregates. Furthermore, earthworms also help to produce large aggregates. Finally, deep-rooted trees help to stabilize steep slopes by increasing the total shear strength of the soils.

It is therefore to be expected that with the removal of forest, for agriculture or for other reasons, rates of soil loss will rise and mass movements will increase in magnitude and frequency. The rates of erosion that result will be particularly high if the ground is left bare; under crops the increase will be less marked. Furthermore, the method of ploughing, the time of planting, the nature of the crop, and size of the fields, will all have an influence on the severity of erosion.

It is seldom that we have reliable records of rates of erosion over a sufficiently long time-span to show just how much human activities have accelerated these effects. Recently, however, techniques have been developed which enable rates of erosion on slopes to be gauged over a lengthy time-span by means of dendrochronological techniques that date the time of root exposure for suitable species of tree (Carrara and Carroll, 1979).

Another way of obtaining long-term trends of soil erosion rate is to look at rates of sedimentation on continental shelves and on lake floors. The former method was employed by Milliman et al. (1987) to evaluate sediment removal down the Yellow River in China during the Holocene. They found that, because of accelerated erosion, rates of sediment accumulation on the shelf over the last 2300 years have been ten times higher than those for the rest of the Holocene (i.e. since around 10,000 BP).

Another good example of using long-term sedimentation rates to infer long-term rates of erosion is provided by Hughes et al.'s (1991) study of the Kuk Swamp in Papua New Guinea (figure 3.14). They identify low rates of erosion until 9000 BP, when, with the onset of the first phase of forest clearance, erosion rates increased from 0.15 cm 1000 a^{-1} to about 1.2 cm 1000 a^{-1}. Rates remained relatively stable until the last few decades when, following European contact, the extension of anthropogenic grasslands, subsistence gardens and coffee plantations has produced a rate that is very markedly higher – 34 cm 1000 a^{-1}.

In the Brecon Beacons of Wales, Jones et al. (1985) have obtained a handle on rates of erosion by calculating rates of sedimentation in Llangorse Lake during the Holocene. They found the following trends:

Period (years BP)	Sedimentation rate (mm 1000a^{-1})
9000 – 7500	350
7500 – 5000	100
5000 – 2800	1320
2800 – AD 1840	1410
c.AD 184 – present	5900

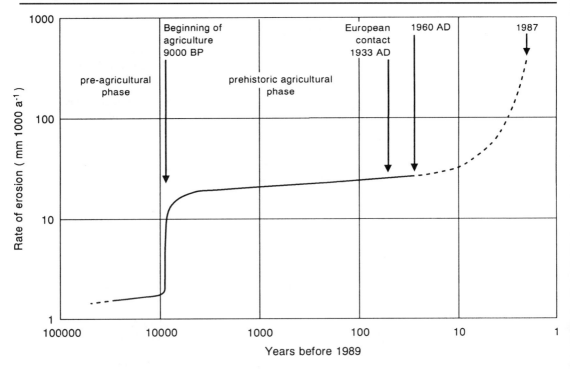

Figure 3.14
*Rates of erosion in mm 1000 a⁻¹ in Papua New Guinea in the Holocene, derived from rates of
sedimentation in Kuk Swamp (source: modified from Hughes et al., 1991, fig. 5).*

The thirteen-fold increase in rates after 5000 BP seems to have
occurred rapidly and is attributed to initial Neolithic forest clearance. The second dramatic increase of more than four-fold took
place in the last 150 years and is a result of agricultural intensi-
fication.

In Italy, Judson (1968) used sediment yield, archaeological and
sediment core data to estimate long-term rates of erosion near
Rome. He suggests that before human settlement of the area in
Classical Times, rates of denudation were around 20 to 30 mm
$1000 \ a^{-1}$. In the last 2000 years rates have been very markedly
higher, commonly ranging between 200 and 1000 mm $1000 \ a^{-1}$,
an increase of over ten-fold.

Dearing and Foster (1993) have tried to bring together some of
the available information on Holocene records of sedimentation
derived from the study of lake deposits (figure 3.15) from different
global regions. The records in figure 3.15a have been generalized for
north-west Europe and compare trends in erosion between landscapes where the human impact has been large and those where it
has been less substantial. In the latter case the highest erosion
rates occurred during the periglacial phase at the end of the lateglacial period. Rates declined thereafter as the landscape became
vegetated in the improving interglacial conditions of the Holocene.

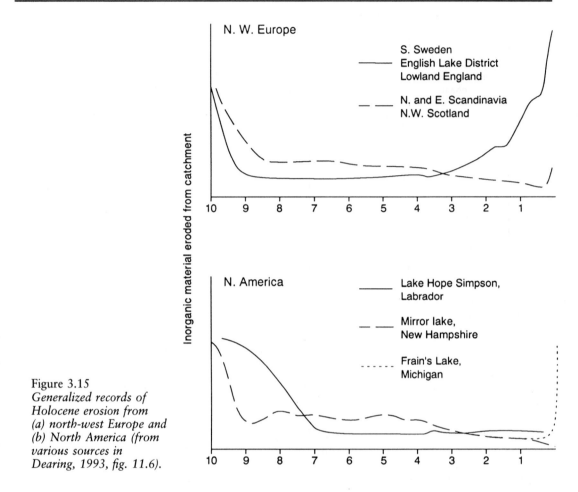

Figure 3.15
*Generalized records of
Holocene erosion from
(a) north-west Europe and
(b) North America (from
various sources in
Dearing, 1993, fig. 11.6).*

Throughout the past 9000–8000 years erosion rates have tended
to remain approximately constant or even to show some decline,
except in the past couple of centuries when more erosion has been
caused by road construction, drainage for afforestation and other
such activities. In contrast Dearing points to the former case where
the landscape has a long history of agriculture and where erosion
rates have generally increased since initial deforestation, some 5000
to 2000 years ago. The rate has tended to accelerate in a series of
jerks as new technologies and land-use practices have been introduced. In North America (figure 3.15b) agricultural history has a
shorter duration for the most part and records from three lakes
suggest that before pioneer agricultural colonization the long-term
rate of erosion was similar to that in north-west Europe. In cooler
and more northerly Labrador the high early rates lasted longer
than in the warmer and more southerly New Hampshire. Spectacular increases in erosion may have followed recent deforestation.

Urbanization is another land-use change that can lead to substantial changes in sediment yields (see table 1.3c).

The highest rates of erosion are produced in the construction phase, when there is a large amount of exposed ground and much disturbance produced by vehicle movements and excavations. Wolman and Schick (1967) and Wolman (1967) have shown that the equivalent of many decades of natural or even agricultural erosion may take place during a single year in areas cleared for construction. In Maryland they found that sediment yields during construction reach 55,000 t km^{-2} a^{-1}, while in the same area rates under forest were around 80–200 t km^{-2} a^{-1} and those under farm 400 t km^{-2} a^{-1}. New road cuttings in Georgia were found to have sediment yields up to 20,000–50,000 t km^{-2} a^{-1}. Likewise, in Devon, England, Walling and Gregory (1970) found that suspended sediment concentrations in streams draining construction areas were two to ten times (occasionally up to 100 times) higher than those in undisturbed areas. In Virginia, USA, Vice et al. (1969) noted equally high rates of erosion during construction and reported that they were ten times those from agricultural land, 200 times those from grassland and 2000 times those from forest in the same area.

However, construction does not go on for ever, and once the disturbance ceases, roads are surfaced, and gardens and lawns are cultivated. The rates of erosion fall dramatically and may be of the same order as those under natural or pre-agricultural conditions. Moreover, even during the construction phase several techniques can be used to reduce sediment removal, including the excavation of settling ponds, the seeding and mulching of bare surfaces, and the erection of rock dams and straw bales.

Although accelerated erosion means that the sediment yields of many rivers may have increased as a result of human activities, it is also important to recognize that in terms of the delivery of that sediment to the oceans many rivers have now been dammed and that as a consequence their sediment loads have been very greatly reduced. The River Nile, for example, now only transports about 8 per cent of its natural load below the Aswan Dam. Even more dramatic is the picture for the Colorado River in the USA. Prior to 1930 it carried around 125–150 million tons of suspended sediment to its delta at the head of the Gulf of California. Following the construction of a series of dams, the Colorado (figure 3.16) now discharges neither sediment nor water into the ocean (Schwarz et al., 1991). Similarly, rivers on the eastern seaboard of the United States draining into the Gulf of Mexico or the Atlantic have shown marked falls in sediment loadings (figure 3.17), and four major Texan rivers (table 3.4) carried, in 1961–70 an average only about one-fifth of what they carried in 1931–40. Likewise, in France the Rhône only carries about 5 per cent of the load that it did in the nineteenth century, while in Asia the Indus discharges less than 20

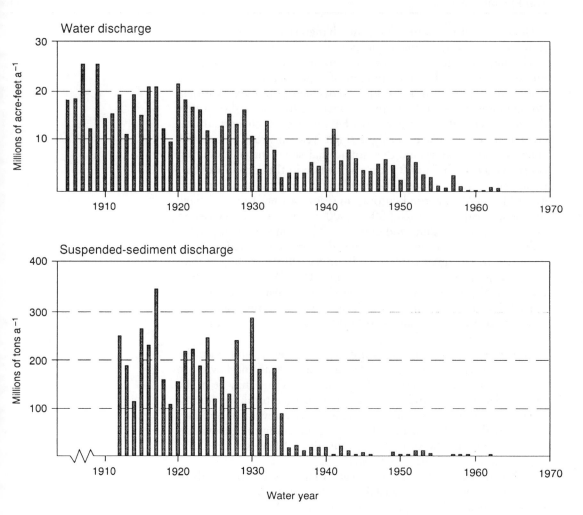

Figure 3.16
Historical sediment and water discharge trends for the Colorado River, USA (source: modified after the United States Geological Survey in Schwarz et al., 1991).

Table 3.4 Suspended loads of Texas rivers discharging into the Gulf of Mexico (million tonnes)

River	1931–40	1961–70	%*
Brazos	350	120	30
San Bernard	1	1	100
Colorado	100	11	10
Rio Grande	180	6	3
Total	631	138	20

*1961–70 loads as % of 1931–40 loads.
Source: Goudie (1990) table 6.14.

per cent of the load it did before the construction of large barrages over the last half century (Milliman, 1990).

In spite of increasing tendency for the delivery of sediment to the oceans by rivers to decline as more and larger dams are constructed, on a global basis it is probable that rates of sediment delivery are still markedly higher than they were in pre-human times. There are plainly huge difficulties involved in estimating erosion rates in pre-human times, but in a recent analysis McLennan (1993) has estimated that the pre-human suspended sediment discharge from the continents to the oceans was about 12.6×10^{15} g a^{-1}, which is about 0.6 of the present figure.

Rates of change within catchments: river bank erosion

Up to this point we have largely been concerned with the gross rates at which rivers have been transporting sediment and denuding the land surface of the Earth. We conclude the chapter by looking at the rates at which certain components of fluvial systems change. The first of these is bank erosion.

Howard (1992) has attempted to summarize the four types of constraint that may limit rates of river bank erosion:

1. The rate of deposition of the point bar.
2. The ability of the stream to remove the bed-load component of the sediment eroded from the bank deposits.
3. The ability of the stream to entrain sediment from *in situ* or mass-wasted bank deposits.
4. The rate at which weathering acts to diminish bank sediment cohesion to the point that particles may be entrained by the flow or bank slumping may occur.

He believes that in general the first of these constraints is not a major limiting factor in most meandering steams, given that banks are generally more cohesive than beds. He recognizes that exceptions could occur, however, if the banks were to be composed of non-cohesive materials that were finer in grain size than the channel bed. The three other constraints relate to the power which the river possesses in relation to the resistance of the bank materials.

Using an international data set, Hooke (1980) was able to present a good summary of available data on rates of channel erosion. She found (figure 3.18) a broad relationship between rate of bank erosion for meandering channels and the drainage area of the river moulding the channel. There is a considerable scatter in this relationship, but it is evident that very large streams can display rates of bank erosion of tens or even hundreds of metres per year, with particularly high rates being recorded for rivers such as the Brahmaputra and the Mississippi. Coleman (1969) demonstrated from cartographic evidence that the former river had moved at

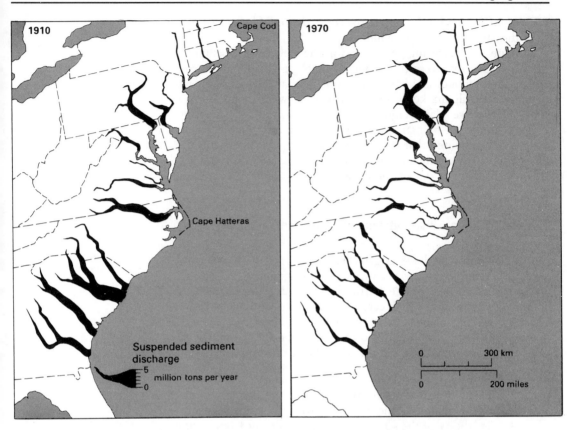

Figure 3.17
The decline in suspended sediment discharge to the eastern seaboard of the United States between 1910 and 1970 as a result of soil conservation measures, dam construction and land-use changes (source: Meade and Trimble, 1974, with modifications).

rates as high as almost 800 metres per year between 1944 and 1952, and that some stretches had migrated as much as 96 km in just 200 years. He stressed the importance of high discharge, high sediment loads, and tectonic instability in enabling such high rates. Brunsden and Kesel (1973) have presented data that suggest that the Mississippi in Louisiana has eroded portions of its banks at rates of up to 23 m a^{-1}. However, some of the very fastest rates of channel change are recorded for the rivers that flow into the great plains of India and Pakistan. The tributaries of both the Indus and the Ganga systems have shown remarkable fluctuations in position and relative importance over historical times. The Kosi River, which flows from the Nepalese Mountains into India and the Ganga shows a lateral shift rate of 112 km in 228 years, which gives a long-term average of about 90 m a^{-1} (figure 3.19). Such rivers are in effect huge humid fans – hence their instability.

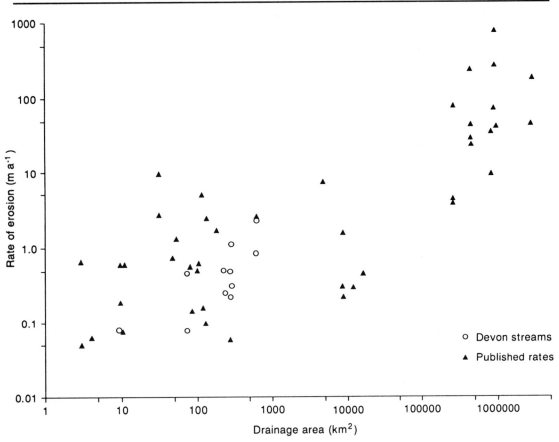

Figure 3.18
*Relationship between rates of river bank erosion and catchment area for streams in Devon, England,
and for other parts of the world (source: Hooke, 1980, fig. 2).*

River bank erosion and channel change in the UK

Hooke (1980) and Newson (1986) provide data on rates of bank
erosion for England and Wales; these are listed in table 3.5. Several things can be said about them. First, bank erosion rate, as
might be expected, is closely related to catchment area (a surro-
gate of discharge and width). Secondly, rates of bank erosion are
such that this may well be an important source of stream suspended loads. Thirdly, because river channel movement is one of
the main processes by which the formation, development and
renewal of floodplains takes place, these data have important
implications for rates for floodplain development. Hooke calcu-
lates that for rivers in Devon (using measured floodplain widths
and mean rates of floodplain development), it takes between around

Table 3.5 Rates of river bank erosion

River	Dates of surveys	Rate (m a⁻¹)
Cound, Shropshire	1972–4	0.64
Bollin, Cheshire	1872–1935	0.16
Exe, Devon	1840–1975	2.58
Creedy, Devon	1840–1975	0.52
Culm, Devon	1840–1975	0.51
Axe, Devon	1840–1975	1.00
Yarty, Devon	1840–1975	1.38
Coly, Devon	1840–1975	0.48
Hookamoor, Devon	1840–1975	0.19
Severn	1948–75	0.2–0.7
Rheidol	1951–71	1.75
Tywi	1905–71	2.65

Source: Newson (1986) table II; Hooke (1980) table IV.

6000 and 7000 years for a complete traverse of the floodplain to take place. Fourthly, the data indicate that at some sites rates are sufficiently high as to cause considerable practical problems for riparian owners.

However, as Ferguson (1981) points out, most British rivers have rather inactive channels and have shown little change since their courses were surveyed accurately for the first time in the late nineteenth century. Some rivers are constrained in rock-walled channels (e.g. the Wye in the Welsh borderland, and the Dove and the Wye in Derbyshire). Others have banks which are protected against erosion by the presence of trees. However, the biggest class of inactive channel in Britain is the clay lowland rivers of East Anglia and the Thames Basin, where low regional slopes, a small rainfall surplus over evaporation and cohesive banks, derived from soft bedrock or Pleistocene drifts, combine to dampen down present-day channel migration.

Although channel changes are ultimately caused by river erosion and transport, detailed monitoring of meander beds in South Wales by Lawler (1986), using grid networks of erosion pins, has highlighted the importance of frost action in preparing banks for subsequent fluvial scour at times of high flow. He found that minimal fluvial erosion occurred on banks which had not previously been prepared by frost attack and that most bank retreat occurred in the winter months of December, January and February. Such preparatory processes are probably particularly important in the upper courses of streams, while as one moves down stream the relative importance of fluid entrainment and mass failure may become greater as, respectively, stream power and bank height become greater.

Table 3.6 Riverbed degradation below dams

River	Dam	Amount (m)	Length (km)	Time (years)
South Canadian (USA)	Conchos	3.1	30	10
Middle Loup (USA)	Milburn	2.3	8	11
Colorado (USA)	Hoover	7.1	111	14
Colorado (USA)	Davis	6.1	52	30
Red (USA)	Denison	2.0	2.8	3
Cheyenne (USA)	Angostura	1.5	8	16
Saalach (Austria)	Reichenhall	3.1	9	21
South Saskatchewan Canada	Diefenbaker	2.4	8	12
Yellow (China)	Sanmexia	4.0	68	4

Source: from data in Galay (1983).

Some anthropogenic channel changes

River channels, because they are relatively dynamic features, have changed markedly in the face of human influences on their water and sediment supply, and on their bank cohesion.

The building of dams can lead to channel aggradation upstream from the reservoir and channel deepening downstream because of the changes brought about in sediment loads. Some data on observed rates of degradation below dams are presented in table 3.6. They show that the average rate of degradation has been of the order of a few metres over a few decades following closure of the dams. However, over a period of time the rate of degradation seems to become less or to cease altogether, and Leopold et al. (1964, p. 455) suggest that this can be brought about in several ways. First, because degradation results in a flattening of the channel slope in the vicinity of the dam – the slope may become so flat that the necessary force to transport the available materials is no longer provided by the flow. Second, the reduction of flood peaks by the dam reduces the competence of the transporting stream to carry some of the material on its bed. Thus if the bed contains a mixture of particle sizes the river may be able to transport the finer sizes but not the larger, and the gradual winnowing of the fine particles will leave an armour of coarser material that prevents further degradation.

The overall effect of the creation of a reservoir by the construction of a dam is to lead to a reduction in downstream channel capacity (see Petts, 1979 for a review). This seems to amount to between about 30 and 70 per cent (see table 3.7).

Other significant changes produced in channels include those prompted by accelerated sedimentation associated with changes in the vegetation communities growing along channels. The introduction of the salt cedar in the southern USA has caused significant floodplain aggradation. In the case of the Brazos River in Texas, for example, the plants encouraged sedimentation by their

Table 3.7 Channel capacity reduction below reservoirs

River	Dam	% channel capacity loss
Republican, USA	Harlan County	66
Arkansas, USA	John Martin	50
Rio Grande, USA	Elephant Buttre	50
Tone, UK	Clatworthy	54
Meavy, UK	Burrator	73
Nidd, UK	Angram	60
Burn, UK	Burn	34
Derwent, UK	Ladybower	40

Source: modified after Petts (1979) table 1.

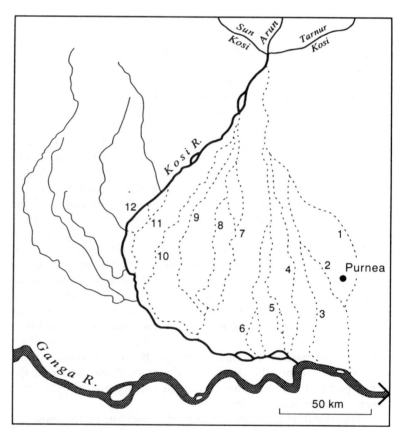

Figure 3.19
Lateral shifts in the course of the Kosi River, north India, in historical times (source: modified from Holmes, 1965):
(1) before 1736, (2) 1736, (3) 1770, (4) 1807–39, (5) 1840–73, (6) 1873–93, (7) 1893–1921, (8) 1922–26, (9) 1926, (10) 1933, (11) 1938–40, (12) 1949.

damming and ponding effect. They clogged channels by invading sand banks and sand bars, and so increased the area subject to flooding. Between 1941 and 1979 the channel width declined from 157 to 67 m, and the amount of aggradation was as much as 5.5 m (Blackburn et al., 1983). Another example of rapid aggradation is that produced by the introduction of mining waste into channels. In mid-Wales the Afon Ystwyth suffered 3 m of aggradation for this reason (Lewin et al., 1983).

Floodplain accretion rates

The Holocene stratigraphy of rivers provides plenty of evidence for the rate at which cut and fill can take place. This can be supplemented by direct measurement of the rates at which sediment accumulates on sediment traps located on floodplain surfaces, or by the use of ^{137}Cs profiling (Walling et al., 1992).

As one might anticipate, rates are highly variable in both space and time, as related to local topographic position and to the nature of individual flood events. Bridge and Leeder (1979, table 1) provide a list of rates of accretion that were available at that time. Their average rates ranged from virtually nothing to as much as 91 cm a^{-1}.

Undoubtedly, many of the world's rivers cut down in response to low glacio-eustatic sea levels in the late Pleistocene, and their late Pleistocene channels have since become aggraded by Holocene deposition. As a rough rule of thumb perhaps 100 m of deposition have occurred in the last 10,000 years, giving a mean rate for the whole Holocene of 10,000 mm 1000 a^{-1}. Resarch in the Indus plain of Pakistan suggests that a substantial proportion of this aggradation was achieved relatively early on in the Holocene (Jorgensen et al., 1993, p. 297).

Data on rates of floodplain alluviation by overbank deposition in the UK are relatively sparse (Lambert and Walling, 1987). Using sediment trap data from floodplain surfaces, and by calculating losses in conveyance in channels, they estimated that in the case of the River Culm, Devon, the present average rate was approximately 0.05 cm a^{-1}. This is a very low rate in comparison with those that have been recorded in the last two centuries in the USA (see below), and is also lower than the long-term rates determined for the Holocene for the River Severn of 0.14 cm a^{-1} (Brown, 1987) and for the Warwickshire Avon (Shotton, 1978).

Workers in the USA examining the consequences of European settlement, deforestation and agriculture, have used stratigraphic methods and archival sources to calculate rates. Many valleys have several metres of depositional material that have been laid down in the last two centuries or so. Costa (1975), for example, found floodplains in the upland areas of the Piedmont Province of Maryland with 0.9 to 1.5 m of recent sediments, implying accretion rates as high as 1.6 cm a^{-1}, while Magilligan (1985), working in Wisconsin and Illinois, found average rates of deposition of 1.9 cm a^{-1} for the 120 years prior to 1940, compared with rates of 0.75 cm a^{-1} for the post-1940 period. Pizzuto (1987) established that in the case of the Brandywine Creek in Pennsylvania, post-settlement deposition amounted to 1.6 m, indicating similar rates to those found in the two earlier studies. Finally, Knox (1989), working in south-western Wisconsin and north-western Illinois found that there had been as much as 3 to 4 m of deposition in the lower reaches of main valleys near the Mississippi River, and that rates of accelerated overbank floodplain sedimentation could be as high as 4.0 to 5.0 cm a^{-1}, compared with an average pre-settlement post-glacial floodplain vertical accretion rate of 0.02 cm a^{-1}.

Long-term rates of karstic drainage incision

Although the prime method that has been used to assess rates of limestone denudation has been the chemical analysis of drainage waters (see pp. 59–65), an alternative way of estimating the speed with which drainage incision and landform evolution has occurred is to use isotopic means – primarily the Uranium-Thorium method – to date diagnostic speleothems. The principle behind this method is that many karstic areas have abandoned, high-level cave passages in which former underground streams drained to valley floor levels that are well above the present ones. Thus by isotopic dating of the speleothem deposits in such caves – deposits which could not have formed while the caves were still fully active – the dates when the caves were abandoned by their streams can be assessed. This means that rates of stream downcutting can be obtained by comparing dates with levels, with the vertical distance between dated cave levels indicating the rate of downcutting that has occurred. It is important to remember that these are rates of localized lowering rather than rates of basin-wide denudation.

In Britain a number of studies have employed this technique in the classic areas of Carboniferous Limestone karst. In north-west Yorkshire (Gascoyne et al., 1983) major cave systems have been found to have ages that are beyond the limit of the ^{230}Th/^{234}U method (i.e. greater than 350,000 years). Maximum downcutting rates of channel beds in the Craven area work out at about 20–50 mm 1000 a^{-1}.

Dates in the Derbyshire Peak District have also revealed that caves have ages greater than 350,000 years (Ford et al., 1983), and Rowe (1988) has proposed, on the basis of not only uranium series methods but also palaeomagnetic techniques, that some of the speleothems may be late Pliocene in age (i.e. 1.8 to 2.0 million years old). Estimated rates of channel incision are rather greater than those quoted for Craven, being 55 mm 1000 a^{-1} for the Manifold valley, and 63 mm 1000 a^{-1} for the Cresswell Crags gorge. Further south, in the Mendips, beyond the normally accepted glacial limit, the karst scenery was also well developed beyond 350,000 years (Atkinson et al., 1978), and long-term maximum rates of channel downcutting have been of the order of mm 1000 a^{-1}.

The abrasion of bed-load material

It is a common observation that river bed-load materials are generally rounded, and that the degree of rounding of particles tends to increase with distance of transport. Such rounding is achieved by abrasion, a process which may also contribute to the frequently observed diminution of clasts downstream. Calculations of the rate at which rounding and size diminution are achieved by abrasion have been carried out on the basis of field observations along stream channels, and also by means of laboratory-based abrasion studies using revolving drums.

Kuenen (1956) attempted to use laboratory studies to ascertain some of the main controlling factors in determining abrasion rates. Among his major findings were that clast abrasion of pebble-sized material is greater on a pebbly floor than a sandy one, and that size of clast is a crucial consideration. Coarse gravel, he found, can lose four to five times as much in percentage terms as does fine gravel at equal velocities of movement.

Sneed and Folk (1958) examined pebbles along the lower Colorado River in Texas, and were particularly concerned with how the rate of abrasion and rounding varied with lithology. In their field of study, they found that nearly all rounding of limestone pebbles was achieved in less than 16 km of transport, whereas chert pebbles only became rounded after transport for 300 km or more.

However, as Kukal (1990, p. 87) has pointed out, there is frequently a discrepancy between results of laboratory experiments and those gained from field observations. Field experiments, such as those of Sneed and Folk, indicate faster rates than might be expected on the basis of laboratory experiments, and the laboratory studies have a general tendency to indicate that abrasion of material in rivers is a very slow process and that pebbles should remain unrounded, unstable and undistintegrated for a long time. Berthois and Portier (1957) found a per kilometer loss for medium subangular sand-sized quartz grains of approximately 0.0025 per cent, indicating that for a mere 1 per cent weight loss to occur approximately 400 km of fluvial transport would be necessary. Likewise Kuenen's (1959) experiment found mechanized abrasion was negligible on material with a diameter between 0.4 mm and 10 mm. The weight loss per kilometer was only 0.00005 per cent, indicating that around 20,000 km of transport would be necessary to produce a 1 per cent weight loss! Such values are plainly unrealistic. The explanation for the discrepancy may lie in the behaviour of pebbles on a river bed. They do not simply roll downstream. They also tend to oscillate in position, rubbing against each other and so being abraded and rounded almost *in situ*. In other words, pebbles in a stream cover a much larger distance than would correspond to mere down-channel movement (Schumm, 1973). Much also depends on the energy conditions simulated in the experiments and the mix of grain sizes employed. Thus the simulation of Wright and Smith (1993), using a fast tumbling mill and a mix of gravel and sand sized materials, caused rapid comminution of the sand size material and the production of large quantities of silt.

Conclusion

Geomorphologists have developed a plethora of techniques to estimate the amount of work that rivers achieve in terms of denudation and sedimentation. Some of these techniques (e.g. river sediment load data) are applicable to short time-spans, some (e.g. reservoir sedimentation rates) are applicable to rather longer time-spans, while others are much more useful to determine rates over millions of years.

Considerable efforts have been made to identify patterns of denudation at the global and continental scales. Climate has always been given special attention as a possible controlling variable, but different studies have identified different climatic zones as having the highest rates of denudation. It is highly unlikely that any one single climatic zone will display uniquely high rates. Indeed there may be a range of climatic zones (e.g. glacial, wet periglacial, semi-arid, monsoonal) which have a tendency to display high rates of fluvial denudation. Plainly vegetation cover, which may be related to climatic conditions, is a major control of the facility with which sediment is moved, but any simple picture of climatic control will be clouded by the operation of other controls which include the tectonic environment, altitude, available relief, and the availability of erodible material (e.g. loess, glacial drift, etc.). In global terms much of the sediment entering the oceans comes from the monsoonal, tectonically active, river basins of south and south east Asia and Oceania, where until recently the contribution of small catchments has been largely ignored. This is in stark contrast to Australia, where, over geological time-spans, rates of denudation have, over enormous tracts of country, been remarkably low.

Available relief is plainly a major control of rates of denudation, with mountainous areas providing much higher sediment loads than plains. Thus as relief becomes reduced through time so will the quantity and size of the sediment load derived form the drainage system.

Sediment yield may also be scale dependent. Most studies suggest that yields tend to be greater per unit area for small basins than for large. This is normally seen as a result of small catchments tending to have steeper terrain, less opportunity for storage on floodplains, and a greater propensity to the effects of high intensity storm events. The relationship is not, however, invariable and the role of sediment stores and their budget is crucial. Nonetheless, in many basins much of the material liberated by hill-slope erosion may remain stored in the catchment as colluvium or alluvium. Human activities may at the present time determine what proportion of eroded material may be evacuated from a catchment.

Rates of denudation over long time-scales are becoming increasingly available with the development of techniques such as fission track dating of apatites and other minerals. Such techniques enable guesses to be made about the time it might take for the development of low-relief peneplains, but also enable long-term (prehuman) rates to be compared with short-term rates which may have been accelerated because of recent human activities and land-use change. Certainly, studies of lake and swamp sedimentation indicate that rates of denudation have been growing substantially during the course of the Holocene. Humans have also changed the rate of operation of channel processes, including bank erosion and floodplain accretion.

FOUR

Slope processes

Introduction

Slopes comprise the greater part of the landscape and as an integral part of drainage basins they provide water and sediment to streams. In the last three decades they have taken their rightful central place in geomorphology and considerable endeavour has been expended in trying to assess the rates at which slopes evolve and the rates at which different processes operate upon them. Some of this work has been undertaken for essentially 'pure' purposes, but much of it has been done because it has a 'practical' or 'applied' utility.

Processes that involve a transfer of slope-forming materials from higher to lower ground, under the influence of gravity and without the primary assistance of a fluid transport agent, such as air, water or ice, are termed mass movements. These movements can be classified according to their speed and thickness and include one or more of the mechanisms of fall, slide, flow or creep.

Techniques for measuring soil movement and loss

Because of the importance to humans of soil erosion and mass movements on slopes, a wide range of techniques has been developed to measure the wide range of processes involved. These include soil creep, rainsplash, surface wash, and rilling.

As *soil creep* is predominantly a slow process and almost imperceptible, measurement techniques need to be sensitive in order to make accurate assessments of its rate (Statham, 1981).

One class of techniques attempts to measure the rate of surface movement. The simplest of these (which has also been a central means of measuring solifluction rates), is to insert pegs a short distance into the soil surface (plate 4.1), or to place painted rocks on the ground surface. Their changing positions are measured from a fixed benchmark by tape, theodolite or electronic distance meter, but sufficiently precise location is difficult, and in any event

Plate 4.1
A very simple attempt to measure the rate of movement of mudslides by the installation of a line of stakes at Black Ven, Dorset, southern England.

surface markers are prone to disturbance by such agents as animals and humans. Surface movements have also been monitored using measurements of angular rotation achieved by instruments such as the Kirkby T-peg (figure 4.1c). This consists of a steel rod (up to 40 cm in length) with an horizontal cross piece (about 28 cm long) formed into a U-shape. V-shaped mounting plates are fixed to the cross piece, which is also equipped with an adjusting screw. An accurate spirit level is placed on the V-supports, and the peg is then levelled using the adjusting screw. As soil creep proceeds, the rod becomes tilted and the graduated spirit level can be

used to determine the change. Assuming that the angular movement of the bar represents average shear in the soil to the depth
inserted, it is possible to calculate relative movement (x) between
the depth of insertion (y) and the ground surface from the following
equation:

$$x = y \tan \Theta$$

where Θ is the angle of rotation of the T-peg. However, the precise
pivot point about which the rotation takes place is not known, so
that it is impossible to convert the observations into reliable estimates of downslope movement.

The second class of soil creep determination technique attempts
to measure movements in the whole soil profile by making an
insertion of something deformable into the soil. According to its
type the insertion can either be re-excavated to record its new
position, or it can be monitored without being disturbed. The
latter approach is in general preferable, as it permits measurements to be made repeatedly over indefinite timesnaps.

The techniques which require re-excavation are of various types.
One simple approach is to drill a vertical hole in the soil profile
using an auger and a plumbbob, and then to place wire, cable,
marbles, coloured sand, wooden discs, plastic discs, etc. in the
hole and backfill. Thin-walled PVC tubing may also be inserted
vertically into auger holes and their new shape after a period of
time can be frozen by filling then with quick-setting cement or
plaster. To obtain creep measurement with this technique, the soil
profile is carefully excavated and its new configuration measured
from a vertical plumb-line. An alternative method of inserting the
materials for deformation is shown in figure 4.1a.

The Young Pit (Young, 1960) is another whole profile technique that requires excavation. In this case, however, rather than
an auger hole, a soil pit is dug. Short metal pins are inserted into
the wall of the pit along a vertical line related to a stable reference
peg at the base. The pit is then filled in, with care being taken to
make sure that the back fill is as much like the original fill as
possible (figure 4.1e).

Those techniques that do not require re-excavation are based on
measuring the deformation of a flexible tube (e.g. dust extraction
tubing). This is emplaced into an auger hole, and then the deformation is measured with such devices as inclinometers (Hutchinson,

Figure 4.1
Some methods for measuring soil creep on slopes:
(a) The glass bead technique.
(b) The wooden disc technique.
(c) Kirkby 'T' Peg.
(d) The buried cone technique.
(e) The Young Pit.
(f) Flexible soil creep tube.

(a)

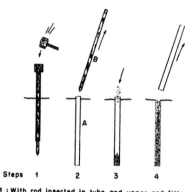

Steps	1	2	3	4

1 : With rod inserted in tube and upper end fitted with driving head, tube and rod are driven into the ground
2 : Withdraw rod
3 : Fill tube with colored grains
4 : Withdraw tube

(b)

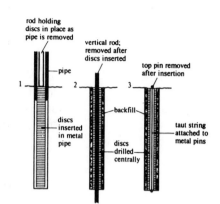

rod holding discs in place as pipe is removed

vertical rod; removed after discs inserted

top pin removed after insertion

pipe

discs inserted in metal pipe

backfill

discs drilled centrally

taut string attached to metal pins

(c)

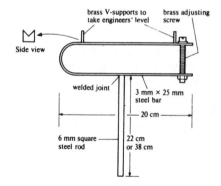

brass V-supports to take engineers' level

brass adjusting screw

Side view

welded joint

3 mm × 25 mm steel bar

20 cm

6 mm square steel rod

22 cm or 38 cm

(d)

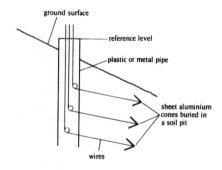

ground surface

reference level

plastic or metal pipe

sheet aluminium cones buried in a soil pit

wires

(e)

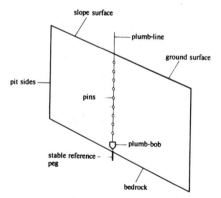

slope surface

plumb-line

ground surface

pit sides

pins

plumb-bob

stable reference peg

bedrock

(f)

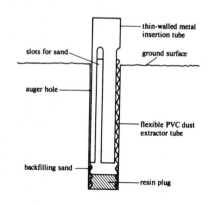

thin-walled metal insertion tube

slots for sand

ground surface

auger hole

flexible PVC dust extractor tube

backfilling sand

resin plug

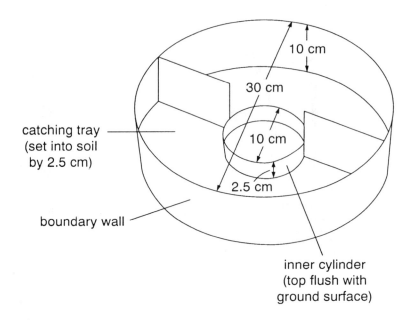

Figure 4.2
Field splash cup (source: from Goudie, 1990, fig. 4.37).

1970) or the travelling telescope (Finlayson and Osmaston, 1977). These devices are relatively complex and expensive, but they do allow relatively precise determinations to be made.

Those techniques that record whole profile movement allow the estimation of the absolute displacement of soil down the slope. The area between successive velocity profiles is calculated and can then be expressed as a volume of displacement per unit width of slope through time (e.g. $cm^3 \ cm^{-1} \ a^{-1}$).

Certain criteria need to be met if *soil splash* erosion is to be measured satisfactorily. The instrument selected must: isolate splash from the effects of sediment movement by overland flow and creep; it must not be affected by relative changes in the height of the device with respect to the soil surface as a result of such processes as ground surface lowering, frost heave, shrinkage and swelling; and it must not interfere with the rainfall properties close to the ground surface. Thus although a variety of techniques has been used (e.g. splash boards; small funnels or bottles inserted in the soil; radioactive tracers; monitoring of painted stones) Morgan (1981) believes that his field splash cup (figure 4.2) is the preferred method. Soil is collected separately from the upslope and downslope compartments of the catching tray, dried and weighed. The combined upslope and downslope weights are a measure of splash detachment, while the downslope weight minus the upslope weight is a measure of the net downslope splash transport. Data may be expressed as unit width measurements (g cm^{-1}) obtained by assuming that the side of a square of the same area as that enclosed by the inner cylinder of the splash cup represents the width across

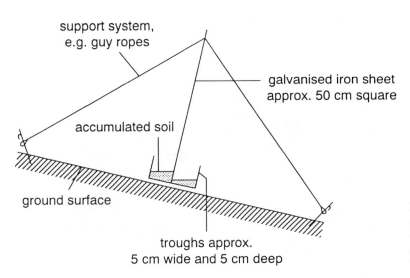

Figure 4.3
An example of a splash board and trough (source: from Goudie, 1981, fig. 4.30).

which splash takes place, or they can be expressed as unit area measurements (kg m^{-2}) based directly on the area enclosed by the inner cylinder.

Other methods that have been employed to measure rain splash include the splashboard technique (Ellison, 1945). A board is erected across the slope with two narrow troughs sunk flush with the soil surface at the base (figure 4.3). These troughs are so positioned either side of the board that they catch material splashed onto the upslope and downslope sides of the board. After a precipitation event splashed material is removed from both sides of the board, and the trough is emptied. The total upslope and downslope catch can then be weighed and net transport rate expressed (as g cm^{-1} of slope). The board technique suffers from the facts that overland flow may enter the troughs, and the boards may influence the turbulence of the wind thereby affecting the amount and intensity of rain on the nearby ground surface.

The most used technique to measure total slopewash erosion, as opposed to rainsplash erosion by itself, is the emplacement of stakes or pins into the ground surface Haigh (1977) provides a good general review of their advantages, problems and pitfalls. Undoubtedly, there are serious soil disturbance problems on insertion and it is advisable to leave them to settle for some time before commencing observations. The pins also influence flow in their vicinity, and their depth in the soil may, under certain circumstances, be influenced by miscellaneous soil heaving processes.

An alternative method is to use an erosion frame. This consists of a rectangular grid of aluminium bars which can be positioned over four permanently installed steel stakes driven into the ground at the measuring site. A depth-gauging rod may be passed through holes drilled into the bars of the frame to measure changes in

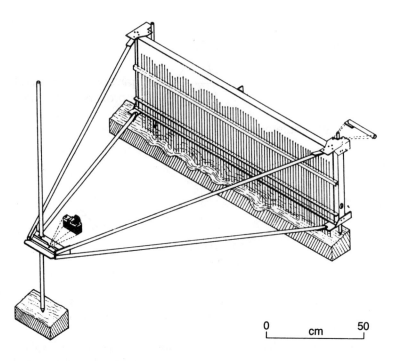

Figure 4.4
*A portable
photographically recording
rill meter (source: McCool
et al., 1981, fig. 1 with
modifications).*

0 _____ cm _____ 50

elevation of the soil surface at a number of points with respect to
the frame. The advantage of this method over erosion pins is that
no disturbance of the soil occurs at the point of measurement
when the instrument is installed, there is no modification of water
flow, and a number of measurements may be made at the same
site (Lam, 1977). Similar considerations prompted Toy (1983) to
develop a Linear Erosion/Elevation Measuring Instrument (LEMI)
based on the use of steel rods placed in the slope together with
two carpenter's levels. It enables the vertical distance between the
LEMI and the soil surface to be measured to the nearest 0.5 mm.
To measure changes in soil rills, various types of rill meter have
been developed (McCool et al., 1981). The example shown in
figure 4.4 is rugged, collapsible and portable and uses a camera as
a recording device, thereby permitting rapid operation. The configuration of the ground surface is shown by long stainless steel
pins that are free to move vertically within a frame.

Marked particles are another method that has been employed to
estimate rates of soil surface movement. The procedure is simple
and inexpensive and allows easy replication of study sites. However,
the method is not without its drawbacks:

1. The tracer particles (e.g. painted stones) may not be truly
   representative of the slope cover on which they are used,
   and since particles of different sizes move at different
   rates, such tracers may predict the movement of only a
   small part of the slope material.

2. Tracer particles tend to be introduced to a site and are
not immediately incorporated into the material whose
movement they are intended to define. As a result, the
marked particles are initially unstable and, during the
period after installation, tend to move downslope more
rapidly than the equivalent particles that are part of the
natural surface. It is thus preferable to mark particles *in
situ*.

The role of organisms in the translocation of sediments on slopes
is one that has not received the attention it deserves, and the
biological factor is normally lacking from the formulae used for
the prediction of soil erosion. Nonetheless, such studies as have
been undertaken on such organisms as isopods and porcupines
(Yair and Rutin 1981), badgers (Voslamber and Veen 1985), worms
(Goudie, 1988), ants and termites indicate their considerable
potential importance. Most studies involve observation of sample
plots and the collection and weighing of casts, mounds, nests,
faeces, etc (see, for example, Hazelhoff et al., 1981).

Total soil removal on a slope from whatever cause can be
achieved by estimating the amount of removal from a sample plot.
This latter approach has certain problems (Roels, 1985). One of
these is that there is no standard plot design. Open plots as well
as bounded ones are used, while the instrumentation involves ranges
from simple wash traps (e.g. Gerlach Troughs) to sophisticated
set-ups using divisors that separate alignotes of runoff during all
the stages of the flow. This is necessary where the volume of
runoff is too great to be collected in a tank. Plot width and length
may distort natural flow patterns, especially when rilling occurs.
There may also be various sources of inaccuracy including leak-
ages into and out of the plot, distortion of the surface at the point
of sediment collection by the existence and implantation of the
measurement devices, blockage or overloading of the collector,
and miscellaneous types of electronic or mechanical malfunction.

In addition there are fundamental sampling problems created by
the variability of soil and terrain properties in space and of runoff
events in time. Representative results ideally demand long monitoring periods and a large number of plots.

Longer-term studies of rates of soil loss on slopes may be achieved
by measuring the degree of root exposure of trees of known age.
For this to be possible, the trees used have to be susceptible to
dendrochronological analysis to determine their age, and they must
also be the type of tree that has a root system that is normally
totally below the surface. Given these conditions, rates of soil loss
over a period of some decades can be established (see, for example, Carrara and Carroll, 1979).

Another long-term method, which may have some potential, is
the use of Caesium-137. This is an artificially generated isotope
which has been produced as a result of thermonuclear weapon
tests (Wise, 1980). It has a half-life of 30 years. Measurement of

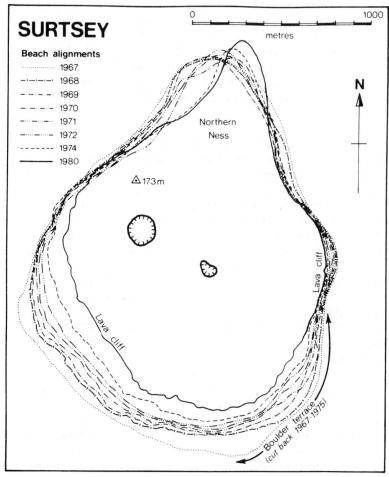

Figure 4.5
The outline of Surtsey, a volcanic island built by eruptions in 1966–7 off southern Iceland, showing the changes revealed by successive air photographs (source: Bird, 1985, fig. 15).

the amount of ^{137}Cs in the soil enables loss or gain to be identified at any point for the period since 1954 (Ritchie et al., 1974), though there are some problems of calibration.

The ratios of ^{137}Cs to [134]Cs can be used to assess erosion rate since the Chernobyl nuclear accident of April 1986 (Branca and Voltaggio, 1993).

Techniques for measuring bank, slope and cliff retreat

There are a variety of techniques that can be employed to measure rates of slope retreat, but the most normal technique for estimating recession distances is to compare the position of a steep slope or cliff on different editions of maps or plans. Other techniques similar to this are to use old and recent terrestrial or air photographs (figure 4.5). Direct measurements using pegs or nails driven into cliffs have also been employed. More recently Lawler (1991) has developed an automatic electronic method for measuring bank retreat, which he calls a Photo-Electronic Erosion Pin (PEEP) system. In some situations it may be possible to use archaeological evidence to obtain longer-term rates (e.g. the loss of Roman settlements in Holderness, England or the undermining of second world war coastal defences). A good general discussion is provided by Sunamura (1983).

One means of determining rates of slopes or cliff retreat is to use what is termed colluviation dating (Statham, 1981, p. 161). The principle of this method is as follows. Landsliding often takes place in well defined phases during which considerable quantities of debris are moved downslope to mantle the formerly exposed lower slopes. The intervening periods of lesser slope activity lead to the spread of vegetation and the formation of soils. Consequently, if the different phases of debris deposition (colluviation) can be dated, possibly from materials included in buried soils, and the volume of material involved in each phase calculated, it should be possible to calculate approximate rates of slope recession over appreciable time-spans.

There are inevitable problems. Material that is suitable for dating has to be found, the estimation of volumes may require extensive boring or pitting, and colluvium on the slope represents only a minimum quantity derived from a higher position on the slope, since sediment and solutes may subsequently have been washed away.

Soil erosion on slopes

The amount of erosion that occurs on a slope as a result of processes such as rainsplash, surface wash and rill action will depend on a range of factors which, primarily, are: the erosivity of the eroding agent, the erodibility of the soil, the slope of the land and the nature of the plant cover.

The loss of soil from a slope is closely related to rainfall characteristics, partly through the power that raindrops have to detach soil particles when they strike the soil surface, and partly through the contribution that rainfall makes to runoff. Various studies have, for example, shown a link between rainfall intensity in any particular storm and the average amount of soil loss that occurs (table 4.1A). However, as Morgan (1986, p. 41) has pointed out, the role of intensity is not invariably so clear, and he draws attention to two main types of rainfall event: the short-lived intense storm which causes the soil's infiltration capacity to be exceeded; and the prolonged storm of lower intensity which thoroughly saturates the soil. Futhermore, antecedent moisture conditions may also determine rates of soil loss (table 4.1B) though even here the relationship is not invariable. Questions also exist over the threshold

Table 4.1 *Soil loss and rainfall conditions for Zanesville, Ohio, USA, for 1934–42*

A   Relationship between rainfall intensity and soil loss

Maximum 5-min intensity (mm h⁻¹)	Number of falls of rain	Average erosion per rainfall (kg m⁻²)
0– 25.4	40	0.37
25.5– 50.8	61	0.60
50.9– 76.2	40	1.18
76.3–101.6	19	1.14
101.7–127.0	13	3.42
127.1–152.4	4	3.63
152.5–177.8	5	3.87
177.9–254.0	1	4.79

B   Influence of antecedent rainfall conditions on soil loss

Date	Rainfall (mm)	Runoff (% of rainfall)	Erosion (g m⁻²)
9 June	19.3	25	1.5
10 June	13.7	66	4.0
11 June	23.8	69	8.9
15 June	14.0	65	4.2
17–18 June	13.0	50	4.6

Data for Zanesville, Ohio, June 1940, for five successive rainstorms on a plot of 20 m²

Source: after Fournier in Morgan (1986) tables 3.1 and 3.3.

values for rainfall that are required to induce significant erosion, but these depend very much on soil properties and the nature of the erosion process. Thus the amount of rain required to cause erosion by overland flow and rilling may be very different from those required to initiate gully formation. Also of great importance is the erosivity of the rainfall, which is a function of a storm's intensity and duration, and of the mass, diameter and velocity of the raindrops.

Soil erodibility varies with soil texture, aggregate stability, shear strength, infiltration capacity and organic and chemical content. For example, large particles are resistant to transport by, for instance, splash, because of the large force required to entrain them, while fine particles can prove resistant to such detachment because of their cohesive qualities. On many slopes in semi-arid regions inverse relationships have been found between stone cover and sediment yield. Such negative correlations can be attributed to several factors, including the fact that stones protect the ground surface soils against aggregate breakdown and surface sealing by raindrop impact, they enhance infiltration rates and thereby diminish runoff, they increase surface runoff which decreases the velocity of overland flow, and they reduce soil detachment and, hence, inter-rill erosion rates. However, the correlation between

Table 4.2   Rates of erosion in selected countries (kg m^{-2} a^{-1})

	Natural	Cultivated	Bare soil
China	< 0.20	15.00–20.00	28.00–36.00
USA	0.003–0.30	0.50–17.00	0.40– 9.00
Ivory Coast	0.003–0.02	0.01– 9.00	1.00–75.00
Nigeria	0.05 –0.10	0.01– 3.50	0.30–15.00
India	0.05 –0.10	0.03– 2.00	1.00– 2.00
Belgium	0.01 –0.05	0.30– 3.00	0.70– 8.20
UK	0.01 –0.05	0.01– 0.30	1.00– 4.50

Source: from numerous sources in Morgan (1986) table 1.1.

sediment yield and percentage stone cover is not always negative (Bunte and Poesen, 1994) and depends on such matters as stone size, the extent to which stones are embedded in the soil and so forth. Thus Poesen and Lavee (1991) indicate that the correlation may become positive where stones are embedded in the soil and are larger than 50 mm. In these circumstances there is an increasing stone-flow effect which outweighs the increasing protection from raindrop impact and flow retardation effects as stone cover increases. Also, the increasing concentration of surface water between the stones leads to increased flow detachment and transport of soil or sediment particles. This is an area that is well reviewed by Abrahams et al. (1994). Also of importance is aggregate stability which depends on humus content, clay amount and size, and factors such as the Exchangeable Sodium Percentage which affect a soil's susceptibility to slaking on wetting. Shear strength, a measure of the soil's cohesiveness and resistance to shearing forces, depends on its physical, chemical and biological properties and on moisture content. Increases in soil moisture content decrease its shear strength and can cause it to deform and flow under its own weight. The soil's infiltration capacity is significant as it tends to determine the amount of surface runoff that may occur and thus cause erosion.

Erosion rates are also related to slope characteristics. It is a broadly reasonable supposition that erosion should increase as slopes become steeper and slope length becomes greater as a result of the respective increases in velocity and volume of surface runoff.

Finally, the nature of the vegetation cover (height, structure and spacing) is highly important in that it acts as a filter between rain and soil. It may reduce raindrop impact, reduce the quantity of precipitation reaching the soil (because of interception and evapotranspirational losses), provide a good soil structure by the provision of humus, and bind the soil through the role of the root system. This is an area where the influence of humans is especially important, and land-use change has a profound effect on erosion rates, as illustrated by table 4.2. It is not, however, only a cover of vascular plants that controls soil erosion. Assemblages of algae (cyanobacteria), lichens, liverworts, mosses and fungi can create cryptogamic or microphytic soil crusts that both protect the soil against raindrop impact and help to produce water-stable aggregates. This may be especially significant in arid and semi-arid areas (Eldridge and Greene, 1994).

Many attempts have been made to combine the various factors that are thought to be crucial in controlling erosion rates into simple predictive models. The most famous of these is the Universal Soil-Loss Equation (USLE) of Wischmeier and Smith (1962). It was designed to estimate soil loss over the long-term from rills and inter-rills at the field or slope scale on moderate slopes and medium soil textures and should not be used to estimate sediment yield from drainage basins or to predict gully or stream bank erosion. The USLE is as follows:

$$E = R.K.L.S.C.P.$$

where E is mean annual soil loss (t ac^{-1} a^{-1}). The derivation of the factors in the equation is as follows.

R. This is the rainfall erosivity index, which is equal to the mean annual erosivity value divided by 100:

$$R = \frac{EI_{30}}{100}$$

K. This is the soil erodibility index defined as mean annual soil loss per unit of erosivity for a standard condition of bare soil, no conservation practice, on a 5° slope, 22 m long. The appropriate value can be obtained from the equation:

$$LS = \frac{\sqrt{1}}{22.13} (0.065 + 0.045S + 0.0065S^2)$$

where L is in m and S in per cent.

C. This is the crop factor. It represents the ratio of soil loss under a given crop to that from bare soil.

P. This is the conservation practice factor. Values are obtained from tables of the ratio of soil loss where contouring and contour strip-cropping are practised to that where they are not. With no conservation measures, the value of P is 1.0. Where terracing is adopted, the value for strip-cropping is used for the P factor, and the LS index is adjusted for the slope length which represents the horizontal spacing between the terraces.

The USLE has no geographic bounds to its use but its application requires knowledge of the local values of its individual factors. Frequently such knowledge is sparse. The model is often applied under circumstances for which it was never designed (Soil Science Society of America, 1979).

Accelerated surface sediment erosion

Humans have accelerated the rates at which surface sediment is removed in a wide variety of ways. The prime causes are deforestation and agriculture, though construction activity, urban growth, war, mining and other such activities are often significant.

The pre-eminent role of deforestation and agriculture was appreciated by early workers like Marsh (1864). Forests protect the underlying soil from the direct effects of rainfall, runoff is generally reduced, tree roots bind the soil, and the litter layer protects the ground from rainsplash. It is therefore to be expected that with the removal of forest, for agriculture or for other reasons, rate of soil loss will rise and mass movements will increase in magnitude and frequency. The rates of erosion that result will be particularly high if the ground is left bare; under crops the increase will be less marked. Furthermore, the method of ploughing, the time of planting, the nature of the crop, and the size of the fields, will all have an influence on the severity of erosion.

It is seldom that we have reliable records of rates of erosion over a sufficiently long time-span to show just how much human activities have accelerated these efforts. Recently, however, techniques have been developed which enable rates of erosion on slopes to be gauged over a lengthy time-span by means of dendrochronological techniques that date the time of root exposure for suitable species of tree. In Colorado, USA, Carrara and Carroll (1979) found that rates over the last 100 years have been about 1800 mm 1000 a^{-1}, whereas in the previous 300 years rates were between around 200 and 500 mm 1000 a^{-1}, indicating an acceleration of about six-fold. This great jump has been attributed to the introduction of large numbers of cattle to the area about a century ago.

Table 4.3, which is based on data from tropical Africa, shows the comparative rates of erosion for three main types of land use: trees, crops and barren soil. It is very evident from these data that under crops, but more especially when ground is left bare, or under fallow, soil erosion rates are greatly magnified. At the same time, and causally related, the percentage of rainfall that becomes runoff is increased.

In some cases the erosion produced by forest removal will be in the form of widespread surface stripping. In other cases the erosion will occur as more spectacular forms of mass movement, such as mudflows, landslides and debris avalanches. Some detailed data on debris-avalanche production in North American catchments as a result of deforestation and forest road construction are presented in table 4.4. They illustrate the substantial effects created by clear cutting and by the construction of logging roads. It is indeed probable that a large proportion of the erosion associated with forestry operations is caused by road construction, and care needs to be exercised to minimize these effects. The digging of drainage

Table 4.3 Runoff and erosion under various covers of vegetation in parts of Africa

Locality	Average annual rainfall (mm)	Slope %	Annual runoff (%)			Erosion (t km^{-2} a^{-1})		
			A	B	C	A	B	C
Ouagadougou (Burkina Faso)	850	0.5	2.5	2–32	40–60	10	60–80	1000–2000
Sefa (Senegal)	1300	1.2	1.0	21.2	39.5	20	730	2130
Bouake (Ivory Coast)	1200	4.0	0.3	0.1–2.6	15–30	10	126	1800–3000
Abidjan (Ivory Coast)	2100	7.0	0.1	0.5–20	38	3	10–9000	10,800–17,000
Mpwapwa* (Tanzania)	c.570	6.0	0.4	26.0	50.4	0	7800	14,600

Note:
A = forest or ungrazed thicket
B = crop
C = Barren soil
* From Rapp et al., (1972), figure 5, p. 259.
Source: after Charreau, table 5.5, p. 153 in Greenland and Lal (1977).

Table 4.4 Debris-avalanche erosion in forest, clear-cut and roaded areas

Site	Period of records (years)	Area		No. of slides	Debris-avalanche erosion (mm 1000 a^{-1})	Rate of debris-avalanche erosion relative to forested areas	
		(%)	(km^2)				
Stequaleho Creek, Olympic Peninsula							
Forest	84	79	19.3	25	71.8	x	1.0
Clear-cut	6	18	4.4	0	0		0
Road	6	3	0.7	83	11,825	x	165
Total	–	–	24.4	108	–	–	
Alder Creek, western Cascade Range, Oregon							
Forest	25	70.5	12.3	7	45.3	x	1.0
Clear-cut	15	26.0	4.5	18	117.1	x	2.6
Road	15	3.5	0.6	75	15,565	x	344
Total	–	–	17.4	100	–	–	
Selected drainages, Coast Mountains, south-west British Columbia							
Forest	32	88.9	246.1	29	11.2	x	1.0
Clear-cut	32	9.5	26.4	18	24.5	x	2.2
Road	32	1.5	4.2	11	282.5	x	25.2
Total	–	–	276.7	58	–	–	–
H. J. Andrews Experimental Forest, western Cascade Range, Oregon							
Forest	25	77.5	49.8	31	35.9	x	1.0
Clear-cut	25	19.3	12.4	30	132.2	x	3.7
Road	25	3.2	2.0	69	1772	x	49
Total	–	–	64.2	130	–	–	–

Source: after D. N. Swanston and F. J. Swanson (1976) table 4.

Table 4.5 Annual rates of soil loss (t km^{-2}) under different land-use types in eastern England

Plot	Splash	Overland flow	Rill	Total
1. *Bare soil*				
Top slope	33	667	10	710
Mid-slope	82	1648	39	1769
Lower-slope	62	1434	6	1502
2. *Bare soil*				
Top slope	60	111		171
Mid-slope	43	778		821
Lower-slope	37	301		338
3. *Grass*				
Top slope	9	9		18
Mid-slope	9	57		68
Lower-slope	12	5		17
4. *Woodland*				
Top slope				0
Mid-slope		1.2		1.2
Lower-slope		0.8		0.8

Source: from Morgan (1977).

ditches in upland pastures and peat moors to permit tree-planting in central Wales has also been found to cause accelerated erosion (Clarke and McCulloch, 1979), while the elevated sediment loads can cause reservoir pollution (Burt et al., 1983).

In general, the greater the deforested proportion of a river basin the higher the sediment yield per unit area will be. In the USA the rate of sediment yield appears to double for every 20 per cent loss in forest cover.

Much will depend, however, on the sources of sediment delivered to river channels. If the bulk of such material is, for example, derived from deep-seated landslides originating well below the root horizon, the removal of trees may have a rather limited effect.

Soil erosion resulting from deforestation and agricultural practice is often thought to be especially serious in tropical areas or semi-arid areas (see T. R. Moore, 1979, for a good case study). However, measurements by Morgan (1977) on sandy soils in the English East Midlands near Bedford indicate the rates of soil loss under bare soil on steep slopes can reach 1769 t km^{-2} a^{-1}, compared with 68 under grass and nothing under woodland (table 4.5).

The use of fire by humans, one of their longest lived technological impacts on the environment, also serves to increase the rate of soil erosion, because fires remove or reduce the vegetation cover, expose the soil to erosive attack, and in some cases change the structure and composition of the soil, thereby rendering them more susceptible to erosive attack.

The burning of forests, for example, can, especially in the first years after the fire event, lead to high rates of soil loss (see table

Table 4.6 Soil losses from burned and protected woodlands in the USA

Location	Forest cover	Years of record	Annual rainfall (mm)	Soil loss (t km² a⁻¹)
Holly Spring (Missouri)	Burned	2	1595	82.5
	Protected	2	1677	26.2
Guthrie (Oklahoma)	Burned	10	765	28.0
	Protected	10	765	2.5
Statesulleo (N. Carolina)	Burned	9	1162	770
	Protected	9	1162	5
Tyler (Texas)	Burned	9	1022	90
	Protected	9	1022	12.5
East Texas	Burned	1.5		51.2
	Protected	1.5		25
North Missouri (A)	Burned	Year 1	1627	127
		Year 2	1010	50
		Year 3	1262	12.5
North Missouri (B)	Protected	Year 1	1627	51.2
		Year 2	1010	25.5
		Year 3	1262	7.5

Source: after data by Ralston and Hatchell (1971), in Foster (1976).

Table 4.7 Soil erosion associated with *Calluna* (heather) burning on the North Yorkshire moors

Condition of Calluna or ground surface	Mean rate of litter accumulation (+) or erosion (–) (mm a⁻¹)	No. of observations
(a) *Calluna* 30–40 cm high Complete canopy	+ 3.81	60
(b) *Calluna* 20–30 cm high Complete canopy	+ 0.25	20
(c) *Calluna* 15–20 cm high 40–100% cover	– 0.74	20
(d) *Calluna* 5–15 cm high 10–100% cover	– 6.4	20
(e) Bare ground. Surface of burnt *Calluna*	– 9.5	19
(f) Bare ground. Surface of peaty or mineral subsoil	– 45.3	25

Source: data in Imeson (1971).

4.6). Burnt forests often have rates a whole order of magnitude higher than those of protected areas. Comparably large changes in soil erosion rates have been observed to result from the burning of heather in the Yorkshire moors in northern England, and the effects of burning may be felt for the six years or more that may be required to regenerate the heather (*Calluna*) (see table 4.7). In the Australian Alps fire in experimental catchments has been found to lead to greatly increased flow in the streams, together with a marked surge in the delivery of suspended load. Combining the two effects of increased flow rate and sediment yield, it was found that after fire the total sediment load was increased 1000 times (Pereira, 1973). Likewise , watershed experiments in the chaparral scrub of Arizona, involving denudation by a destructive fire, indicated that whereas erosion losses prior to the fire were only 43 t km^{-2} a^{-1}, after the fire they were between 50,000 and 150,000 t km^{-2} a^{-1}. The causes of the marked erosion associated with chaparral burning are particularly interesting. There is normally a distinctive 'non-wettable' layer in the soils supporting chaparral. This layer, composed of soil particles coated by hydrophobic substances leached from the shrubs or their litter, is normally associated with the upper part of the soil profile (Mooney and Parsons, 1973), and builds up through time in the unburned chaparral. The high temperatures which accompany chaparral fires cause these hydrophobic substances to be distilled so that they condense on lower soil layers. This process results in a shallow layer of wettable soil overlying a non-wettable layer. Such a condition, especially on steep slopes, can result in severe surface erosion.

Mass movements on slopes

During the last three decades a great deal of effort has been expended on measuring the rates at which the various forms of mass movement process operate. Figure 4.6 gives a general indication of the substantial range that exists both within and between process types. We shall now discuss some of the individual processes in greater detail.

Soil creep

Young and Saunders (1986) gathered together much of the available information that was then in the literature on the rate of operation of different slope processes. They suggested that measurements of soil creep in temperate maritime areas had typically produced values of between 0.5 and 2.0 mm a^{-1} (500–2000 mm 1000 a^{-1}), whereas in temperate continental climates, 'probably because of the more severe ground feezing in winter' (p. 61) values typically ranged up to 15 mm a^{-1} (15,000 mm 1000 a^{-1}). They indicated that there were insufficient data for comparative purposes to enable an assessment of rates in the humid tropical zone.

They investigated the possible role of other controlling factors and suggested, for example, that clay rich soils were likely to show faster rates than sandy soils, and that moisture content would be of considerable importance. Indeed (p. 8) they reported that 'Several studies have found that wet sites move much faster than dry, a fact which destroys the relation of rate with slope angle. This means, too, that creep may remain as fast on gently sloping concavities as higher up the slope, as these become waterlogged more often.'

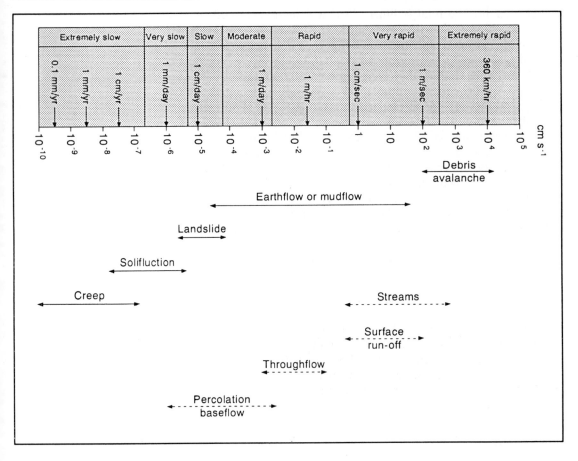

Figure 4.6
General ranges of transport velocities that operate on slopes (source: Clark and Small, 1982, fig. 3.13).

The lack of a clear relationship between slope angle and rate of soil creep has been noted in three separate studies from Great Britain (Finlayson, 1981, in the Mendips; Anderson and Cox, 1984, in upper Weardale, and Slaymaker, 1972, in the upper Wye valley).

Young and Saunders (1986, p. 8) also attempted to evaluate the general landforming significance of this process and remarked, 'The inescapable conclusion is that soil creep can rarely be the predominant denudational process on slopes, by which is meant the process which causes more slope retreat than any other. Only on substantially curved convexities or in seepage zones might this be the case.'

One of the longest and therefore most reliable and useful records of soil creep is that provided by Jahn (1989) for the Sudetes Mountains of Poland. He measured creep for 30 years by means of wooden peg columns inserted in the soil. He found that there was an altitudinal zonation of rates, with negligible values being obtained for the zone of old forest (800–1200 m altitude), values of 2 to 3 mm a^{-1} (2000–3000 mm 1000 a^{-1}) for low meadows (< 800 m) and much higher values of 8–9 mm a^{-1} (8000–9000 mm 1000 a^{-1}) for high meadows (above 1200 m). The low rates of creep in the forests were attributed to their possible role in binding the soil through their roots, but more particularly to the fact that under forest there were far less temperature and moisture variations to drive the creep process. Jahn also looked at the role of slope angle and found that creep tended to occur at greater rates as gradient increased on dry sites, but that with wet sites the relationship was much less clear. This is partly because as sites get wetter the density of the vegetation increases, thereby reducing the rate of soil creep.

Solifluction and gelifluction

Solifluction is a term first used by Anderson (1906) to describe the 'slow flowing from higher to lower ground of waste saturated with water'. He observed the process in the Falkland Islands and the process has often been regarded as characteristic of, though not restricted to, periglacial environments. It is a form of viscous flow that produces a range of features which include sheets, tongue-shaped lobes and alternating stripes of coarse and fine sediment. When the process operates in the active layer of areas underlain by permafrost it is termed gelifluction.

One of the longest term studies of rates of solifluction is that of Smith (1992) who used inclinometer tubes to monitor rates in the Canadian Rocky Mountains over a ten-year period. He found that it took three to four years for the tubes to stabilize, which suggests that studies using this technique for very short periods may tend to over-estimate rates very seriously. Furthermore, analysis of the spatial sampling strategies reported in the literature shows a wide variety of designs that may reduce the present comparability of annual average movement rates between different areas. So, for example, if more movement markers are situated in lateral areas of a solifluction lobe than in central areas, then average rates of movement will be lower than if more markers were located on central areas of the upper lobe surfaces.

Solifluction is a process which operates at a faster rate than does soil creep, and the majority of measured values cluster in the 10,000–100,000 mm 1000 a^{-1} range (Young and Saunders, 1986, p. 8). Once again, however, soil moisture content appears to be a critical control of rate of operation, being a fundamental component of both flow and frost heave. Silty soils are favoured over clays, which are too cohesive, and sands, which drain too easily.

There have been various attempts to try and relate observed rates of solifluction to environmental controls, though in general the length of observations of solifluction itself and the thoroughness of environmental monitoring during the period of solifluction measurements tend to be inadequate in most cases to allow any very firm conclusions to be drawn. Many of the available data for European sites, primarily in the Alps and in Scandinavia, have been reviewed by Matthews and Berrisford (1993). A very wide range of environmental controls are considered, including gradient, vegetation cover, frequency of freeze-thaw cycles, depth of freezing, rapidity of snow melt, summer rainfall amounts, the presence of snowpatches, winter snowfall totals, and grain-size characteristics of sediment. Solifluction is also highly seasonal in occurrence and long-term observation in Spitzbergen by Åkerman (1993) demonstrates that virtually all movement takes place in the summer months of June, July and August.

Surface wash

Surface wash rates have proved to be extraordinarily variable both in time and space, and among the factors responsible for this diversity are variations in rainfall intensity, the degree and type of vegetation cover, soil erodibility, slope angle, and slope length. Young and Saunders, (1986, p. 9) summarized the results of their examination of available data thus: 'However much one tries to generalize the data, the inescapable fact is that rates of wash vary widely, from 1B (Bubnoff Unit) (the lower limit of measurement accuracy due to ground disturbance) in some temperate deciduous woodlands up to 1000–10,000 mm 1000 a^{-1} on badlands. Leaving these two situations aside, most records cover two orders of magnitude, the range being 2–200 mm 1000 a^{-1}.' They also assert (p. 10): 'There can be little doubt that surface wash is the predominant denudational process in semi-arid and tropical subhumid climates, and probably in deserts too; contrary to early supposition, it is substantial under rain forest as well.'

Debris flows

Debris flows (plate 4.2) are a form of gravity-induced mass movement of a body of granular solids, water and air, and are intermediate in type between landslides and water flows. They include mudflows, lahars and the like. They originate when poorly sorted rock and soil debris are mobilized from hill slopes and channels by the addition of moisture (Costa, 1984).

It has often been remarked that debris flows resemble wet concrete in consistency. This material moves down a slope not in a smooth manner but as a series of waves and surges, for periods that range from a matter of seconds to some hours.

Plate 4.2
Debris flows, such as the examples that run down the steep debris fans of the Karakoram Mountains of Pakistan into the Hunza River, are among the fastest of slope processes.

The velocity at which debris flows move varies because of the character of the debris of which they are composed (size, concentration, sorting, etc.), and because of the geometry of their track (shape, slope, width, sinuosity, etc.). Table 4.8 summarizes some available data, and indicates that observed velocities range from less than about 0.5 m s^{-1} to over 20 m s^{-1}.

It is probable that debris flows can themselves cause considerable erosion of their tracks. Their density means that they can exert high shear stresses, and incision during individual events can be of some metres. They can also transport large clasts (boulders over 3000 tonnes in mass have been reported) over large distances (some tens of km), and the volumes of debris involved can be over 10^6 m^3. On alluvial fans, in which they play a major formative role, average rates of accretion can be considerable, and Costa (1984, p. 286) gives values that range between 150 mm 1000 a^{-1} and over 27,000 mm 1000 a^{-1}.

Table 4.8 Rates of debris flow movement

Location	Velocity (m s⁻¹)	Slope (%)
Rio Reventado, Costa Rica	2.9–10	4.6–17.4
Hunshui Gully, China	10–13	–
Bullock Creek, New Zealand	2.5–5.0	10.5
Pine Creek, Mt. St. Helens, Wa	10–31.1	7–32
Wrightwood Canyon, Ca. (1969 flow)	0.6–3.8	9–31
Wrightwood Canyon, Ca. (1941 flow)	1.2–4.4	9–31
Lesser Almatinka River, former USSR	4.3–11.1	10–18
Matanuska Glacier, Alaska	0.001–1.3	2–47
Nojiri River, Japan	12.7–13.0	5.8–9.2
Mayflower Gulch, Colorado	2.5	27
Dragon Creek, Arizona	7.0	5.9

Source: after various sources summarized in Costa (1984) table 1.

Table 4.9 Summary rates of volcanic related processes

Process	Rate	
	Average (m s⁻¹)	Max. (m s⁻¹)
Lava flow	less than 5	less than 30
Ballistic projectiles	50–100	100+
Tephra falls	less than 15	c.30
Pyroclastic flows and debris avalanche	20–30	less than 100
Lahars and jokulhlaups	3–10	30+

Source: from Blong (1984) table 2.13.

Movements associated with volcanic activity

Volcanic areas provide a very special environment in terms of the movement of materials across the land surface, and the rates at which volcanic materials have been moved are well summarized by Blong (1984) (table 4.9).

The velocity at which lava flows is clearly affected by a range of factors which include lava temperature, viscosity, and yield strength, effusion rate, gradient and topographic position. Also, other factors being constant, flow velocity tends to diminish away from its source. The fastest rates are experienced at the surface and in the middle of a flow, until surface crusting occurs, whereupon rates are fastest beneath the flow surface. In general, silicic lava moves more slowly than basaltic, while pahoehoe flows have higher velocities than aa flows. Recorded velocities presented by Blong range from 3 to 3600 m per hour, but may be even greater in confined channels and chutes. Average velocities are less than 5 m s⁻¹. Pyroclastic flows and debris avalanches move at faster rates, with values that range between 9 and 170 m s⁻¹, and which average between 20 and 30 m s⁻¹. Lahars, on the other hand, are slower than pyroclastic flows, though some examples from the Mount St Helens event achieved rates as high as 140 km hr⁻¹. The

Table 4.10 Rate of production of large volcanic accumulations

Location	Area (km²)	Av. thickness (km)	Volume (km³)	Duration of volcanic activity (Ma)	Rate of production (km³ a⁻¹)
S. Africa (Karroo)	2.10^6	7	$1.4.10^6$	20–100	0/07–0/04
W. Siberian Plateau	$2.5.10^6$	0.36	9.10^5	130	0.007
Parana Plateau (S. America)	$1.2.10^6$	0.65	$7.7.10^5$	30	0.026
Deccan Traps	5.10^5	1	5.10^5	10–20	0.025–0.05
Columbia R. (USA)	$2.2.10^5$	0.9	$1.9.10^5$	10	0.019
E. Greenland	–	–	–	3	0.2
Iceland	–	–	–	–	0.03–0.05

Source: from Kukal (1990) (table 24 and text).

fastest rates of movement are, not surprisingly, achieved by ballistic particles, the average rate for which is between 50 and 100 m s⁻¹.

Lava flows show as great a variability in their rate of operation as do most geomorphological phenomena. Their velocities depend on such factors as lava viscosity, distance from the source, the nature and form of the slope and the nature of the flow. Observations from the East African lava lake of Nyiarongo and the Icelandic Surtsey volcano, indicate that in exceptional circumstances velocities may exceed 60 km hr⁻¹.

Giant lava flows are one of the most impressive manifestations of volcanic activity. Classic areas include the Columbia Plateau in the USA, Iceland, the Deccan of India, and the Karroo lavas of the Jurassic of southern Africa. Table 4.10 gives some long-term lava production rates for such volcanic plateau regions, based on studies of dates and the quantities of lava produced. The rates range from a low value of 0.007 km³ a⁻¹ to 0.2 km³ a⁻¹. The latter of these two extremes comes from the eastern plateau of Greenland, where plateau basalts were extruded at this mean rate over a period of three million years (Kukal, 1990, p. 63).

Avalanches and slope development

The role of avalanches was graphically described by Reclus (1881, pp. 117–18):

Of all the destroyers of the mountains an avalanche is the most energetic. It carries away with it earth and rock fragments as would an overflowing torrent; still more by the gradual melting of the snow, forming its lower strata, it so moistens the ground that the latter becomes changed into soft mud, fissured with deep crevices and sinking down beneath its own weight. The earth has become fluid to a great depth, it flows along the whole length of the slopes, drawing with it footpaths, blocks of scattered rocks, even houses and forests. Whole sides of mountains, rendered sodden by

Plate 4.3
In Alpine areas, such as the southern Alps of New Zealand near Mount Cook and the Tasman Glacier, avalanche activity may be an important cause of both erosion and accretion.

the snow, have thus slipped down in one mass with their fields, their pastures, their woods and their inhabitants. Thus by their heating up, and the melting water penetrating so slowly into the ground, flakes of snow suffice little by little to demolish the mountains. In spring every ravine clearly betrays this work of destruction; cascades, landslips, avalanches, snow, rocks and water descend in confusion from the summits and make their way towards the plain.

In high altitude or high latitude environments with a large proportion of their precipitation falling as snow, avalanches are of frequent occurrence and can cause the translocation of large amounts of material, contributing to both erosion and deposition (plate 4.3). The monitoring of such activity can be achieved by the use of various types of trap or by investigating the quantities of material that accumulate on snow banks (e.g. André, 1990).

With regard to accretion of debris transported by avalanches, studies indicate a wide range of values depending on how dirty the avalanches are. In north-west Spitsbergen, André (1990) found

Table 4.11 *Accretion rates of debris supplied by snow avalanches in the Canadian Rockies and on Spitsbergen*

Process	Study area	Lithology	Observation period	Mean accretion rates (mm 1000 a^{-1})		Author
				Min.Max	Individual maxima	
Snow avalanches	Canadian Rockies (Jasper National Park)	Limestone, shales	1968–76	10–7620	29,420	Luckman (1978)
Snow avalanches debris flows	Canadian Rockies (Mount Rae)	Limestone, sandstone, shales	1975–82	20–6670	40,000	Gardner (1983)
Snow avalanches	Spitsbergen (Kongsfjord)	Mica schist, quartzite, gneiss	1983–85	40–8130	41,000	André (1988)

annual accretion rates that varied from 0.04 to 40 mm, and she pointed to the rather remarkable similarity between her values and those that had been obtained from the Rockies by Luckman (1978) and Gardner (1983) (table 4.11). She also stressed the importance of lithology and structure in determining rates, with chute formation being the crucial link. The size of the chutes is determined by rock type, and the larger and deeper the chutes are the dirtier they become. Thus in Spitsbergen rates of accretion are greater in schist bedrock zones than in the more massive gneisses (André, 1988).

Rather longer-term estimates of the role of avalanches in causing debris accumulation can be achieved by estimating the thickness of material that has accumulated on organic horizons of known age (Nyberg, 1987). In the northern Swedish mountains Nyberg found that during the course of the Holocene slush avalanches had led to debris accumulation at about 500–700 mm 1000 a^{-1}.

Individual avalanche events can transport an exceptional quantity of debris (e.g. Ackroyd, 1986), which in turn means that they can achieve a great deal of denudation if climatic, terrain and debris-supply factors are favourable. The evidence of such denudation includes the serrated mountain fronts and avalanche chutes or couloirs of many mountainous areas. In Spitsbergen, André (1990) estimated that rates of denudation due to avalanche activity on cirque walls varied from 7 mm 1000 a^{-1} on massive gneisses to 80 mm 1000 a^{-1} on fractured mica schists.

There are very few studies that attempt to compare the rates of denudation achieved by avalanches with those achieved in the same location by other processes. However, Rapp's pioneering study did just this (Rapp, 1960) in the Karkevagge area of Arctic Sweden. He found that the two most powerful processes there were the transportation of salts in solution in running water and the movements of sediment by earthslides and mudflows. The third most effective process was avalanche activity (especially dirty slushers), followed by rockfalls, solifluction and talus-creep.

Table 4.12 Summary of movement velocities for active rock glaciers

Location	Velocity (cm a⁻¹)	Reference
Swizerland	131–161 (surface)	Chaix (1943)
Switzerland	25–30	Barsch (1969)
Alaska	49–73	Wahrhaftig and Cox (1959)
Yukon	250	Hughes (1966)
Colorado	68	Bryant (1971)
Colorado	5–10	White (1971)
Colorado	8.4–12.5	Miller (1973)
Wyoming	64	Potter (1967)
Average	74	

Source: from Vitek and Giardino (1987) table 16.2.

Rock glaciers

Rock glaciers, which are comparatively common in many alpine areas, are features which look like glaciers but which are composed for the most part of rock debris rather than ice. One of the most important distinctions between rock glaciers and true glaciers is that rock glaciers generally move only very slowly and at least an order of magnitude less than most true ice glaciers. Values reported in the literature for 'active' rock glaciers range from less than 1 cm a⁻¹ to more than 130 cm a⁻¹ (table 4.12). As discussed in chapter 6, rates of true glacier movement are of the order of 3 to 300 m a⁻¹.

As usual there are problems with comparing rates shown in table 4.12 with regard to one another, for techniques of measurement have varied from site to site, and measurements have been made at different points in and on the rock glaciers.

Rates of free face retreat

An ingenious method of trying to estimate the speed at which free faces can retreat in the face of intense periglacial processes is to measure the volume of protalus rampart material that has accumulated during a specific span of time. This was attempted by Ballantyne and Kirkbride (1987) for British ramparts that formed during the Loch Lomond Stadial (*c.*11,000 to 10,000 BP). Their data, which are summarized in table 4.13, indicate an average amount of Loch Lomond Stadial rockwall retreat of 1.35 m, which implies a retreat rate of 1.69 mm a⁻¹ (if the retreat took 800 years) and of 3.38 if the retreat took place over 400 years. These rates are two orders of magnitude greater than estimates of present-day rockwall retreat in Britain (table 4.14), which have been estimated from the volume of recent rockfall deposits. They appear to be broadly comparable to values obtained from present Alpine environments, but considerably higher than those obtained from arctic

Table 4.13 Rockwall retreat rates during the Loch Lomond Stadial calculated from the volume of protalus ramparts

Area	Rampart	Lithology	Average amount of rockwall retreat (m)	Calculated stadial rockwall retreat rate (mm 1000 a⁻¹)	
				Over 800 yrs	Over 400 yrs
Cairngorms	Lairig Ghru 1	Granite	1.61	2010	4020
Cairngorms	Lairig Ghru 2	Granite	1.32	1650	3300
Cairngorms	Devil's Point	Granite	1.27	1590	3180
S.E. Grampians	Dalmulzie	Schist	1.51	1890	3780
Lake District	Dead Crags	Slate	1.33	1660	3320
Lake District	Herdus Scaw	Basalt	1.14	1430	2860
St Kilda	Conachair	Granophyre	1.35	1690	3380
NW Highlands	An Teallach	Sandstone	1.27	1590	3180
	Mean for all sites		1.35	1690	3380

Source: Ballantyne and Kirkbride (1987) table 2.

Table 4.14 Present-day rockwall retreat rates recorded in Great Britain, arctic and subarctic areas, and alpine environments

Location	Lithology	Rockwall retreat rate (mm 1000 a⁻¹)		
		Min.	Mean	Max.
1. Great Britain				
An Teallach	Sandstone	13	15	16
Lomond Hills	Quartz-dolerite	9	15	63
Snowdonia	Volcanics	10	15	21
2. Arctic and subarctic areas				
Lappland	Schist	40	–	150
Spitsbergen	Limestone	50	–	500
Yukon	Various	7	–	170
Ellesmere Island	Limestone	300	–	1300
Spitsbergen	Quartzite	100	–	720
3. Alpine environments				
Swiss Alps	Various	1000	–	2500
Polish Tatras	Dolomite	1000	–	3000
Colorado Front Range	Various	–	760	–
Austrian Alps	Gneiss, schist	700	–	1000

Source: Ballantyne and Kirkbride (1987) table 3, from various sources.

and subarctic areas. Ballantyne and Kirkbride (1987, p. 89) propose various reasons why the Loch Lomond Stadial may have been such a vigorous environment for slope development:

the mountains of Great Britain may have experienced a high frequency of freeze-thaw cycles during the stadial. Conditions for freeze-thaw activity were probably favoured by a combination of strong insolation during spring, summer and autumn months with much cooler air temperatures than at present, a situation for which mid-latitude alpine mountains offer the closest modern analogue. However, the relative instability of over-steepened rockwalls following ice sheet deglaciation is also likely to have resulted in enhanced rockfall activity during the Lateglacial and probably played some part in influencing the rapid stadial rockwall retreat rates reported here.

Another major morphogenetic environment in which free faces and cliffs are thought to be important landscape components is the arid realm. The data for this particular type of environment have been briefly reviewed by Cole and Mayer (1982), who largely used data from the south-west of the USA. The values, which are determined in a whole suite of different ways, the most ingenious of which is the position of packrat middens of known age, average about 0.4 mm a^{-1}, a lower rate than for the periglacial environment of Loch Lomond in Britain, or for the present alpine environments (table 4.14).

On the other hand the Caprock Escarpment of the Texas High Plains, largely composed of relatively incompetent beds of Pliocene age, appears to have undergone rather speedy retreat, with scarp retreat rates of 11 to 18 cm a^{-1} being calculated (Gustavson and Finley, 1985).

Two primary long-term controls of rates of scarp retreat are rate of base level lowering and the angle of rock dip. Howard (1994, p. 142) suggests, for example, that horizontal retreat of a scarp should be inversely proportional to the tangent of the dip, and that the volume of caprock eroded per unit time would be inversely proportional to the size of the dip. Schmidt (1989), however, who worked on the long-term rates of retreat of the great scarps of the Colorado Plateau and found that differences in the rates of scarp retreat stretched over one order of magnitude (between 0.5 km Ma^{-1} for the Kaibab Limestone Cliffs of the Grand Canyon to 6.7 km Ma^{-1} for the Shinarump Conglomerate Chocolate Cliffs), established that a highly significant correlation existed with caprock thickness and an index of rock resistance (dependant on a number of factors such as jointing, bedding structure, grain size, texture, degree of cementation, and nature of cementation material). Schmidt (1988) also estimated rates of scarp retreat in the dry Anti Atlas of Morocco and reported values of 0.5 km Ma^{-1} for resistant limestones and 1.3 km Ma^{-1} for slopes capped with thin conglomerates.

The role of termites and worms in moving sediment on slopes

Organisms undoubtedly play a very major role in mobilizing sediment (see Viles, 1988; for a review). This is especially true of those

Plate 4.4
The role of organisms, including termites (which produced this 2 m high mound in central Namibia) and worms in translocating material on slopes, is a greatly under-researched area in geomorphology.

organisms that spend most of their life cycles in the soil itself, and which attain substantial numbers. Pre-eminent here are termites and worms (Goudie, 1988).

Lee and Wood (1971a) identify three main ways in which termites can contribute to accelerated rates of soil denudation:

(i)   by removing the plant cover;
(ii) by digesting or removing organic matter which would otherwise be incorporated into the soil, and thus making the soil more susceptible to erosion;
(iii) by bringing to the surface fine grained materials for subsequent wash and creep action.

The huge numbers of termites and their large total biomass in favoured localities ensures that these three mechanisms are important. The live weight biomass of termites can be substantial and is comparable to the live weight biomass of large mammalian herbivores in tropical areas.

Some attempts have been made to quantify the speed with which termite mounds (plate 4.4) are constructed and destroyed. Skaife (1955), found that in the Cape area of South Africa mounds grew at a rate of 25 mm a^{-1} (*Amitermes atlanticus*) and that mound growth ceased after around 25 years. Lepage (1984) noted that *Macrotermes* mounds in the Ivory Coast reached a height of 1 m between 2 to 3 years after their first appearance above ground, and that a 3 m mound will be 8–10 years old. Mounds may then be abandoned and observations by Holt et al. (1980) in Queensland, Australia, suggest that *Amitermes vitiosus* mounds are inhabited for 20 to 40 years. In Senegal Lepage (1974) showed that mounds of *Bellicositermes bellicosus* were occupied for rather longer and continued to grow for at least 75–80 years, by which time they had a volume of 50 m^3. In Australia, Spain et al. (1983) calculated that the generation time of mounds was of the order of 30 years.

It would probably be misleading, however, to give the impression that mounds are always short-lived. As Darlington (1985) has pointed out, once mounds are established they provide the best sites for new nests, so that in some cases they will persist over long periods of the order of centuries by repeated recolonization of the same sites. Mounds can also possess high compressive strengths that may provide them with some resistance to erosion. Schmidt Hammer tests on termite mounds in Nigeria (Adepegba and Adegoke, 1974) showed that they had compressive strengths in kg cm^{-2} of 17.4–48.5 compared with 1.4–3.8 for neighbouring unstabilized soils.

Once the mounds have been abandoned they are subjected to erosion. Lepage (1974) estimated that a 1.75 m high mound can be completely eroded in about 50 years, and that 20 to 25 years is required to erode a 8 m^3 mound (Lepage, 1984), whereas Williams (1968) working in the Northern Territory of Australia, estimated that abandoned mounds of *Tumulitermes hastilis* are eroded to near ground level in three years and those of *Nasutitermes triodiae* in about ten years. As the organic matter incorporated in the mound decays, the aggregates that give the mounds their strength and hardness when occupied are destroyed, and rainwash spreads the clay rich material over the surrounding ground (Tricart, 1972, pp. 191–2). The erosion may be hastened by domestic animals and by wild mammals. Ant bears may dig into termitaria, elephants may eat the mineral rich soils in mounds, while animals like cheetah, lion, antelope and buffalo may use them as observation points. Mounds become surrounded by a wash pediment formed of soil eroded from the termitarium (Pullan, 1979).

However, in addition to the potential for erosion and sediment yield caused by mound formation and abandonment it is important to remember the other major consequence of termite-caused soil translocation. This is the construction of covered runways or 'sheetings' on the ground surface and on vegetation. These are constructed of soil particles cemented together with salivary secretions (Bagine, 1984). Three studies in Africa give an indication of the quantities of material involved in this process: in southern Nigeria, Wood and Sands (1978) calculated a rate of 30.0 t km^{-2} a^{-1}; in Senegal, Lepage (1974) found that *Macrotermes subhyalinus* moved 67.5–90.0 t km^{-2} a^{-1}; while in Kenya Bagine (1984) estimated a rate of 105.9 t km^{-2} a^{-1}.

Although the rates of soil translocation by sheeting formation reported above are significant, they are not as important as casts by worms, which seem on average to operate at a whole order of magnitude faster rate. Watanabe and Ruaysoongnern (1984) show on a global basis that cast production by worms operates at rates

Table 4.15 Estimates of speed at which erosion of termite mounds builds a new soil layer

Location	Source	Rate (mm 1000 a⁻¹)
Nigeria	Nye (1955)	25
Uganda	Pomeroy (1976)	40–115
Ivory Coast	Lepage (1984)	750–1000
Australia	Lee and Wood (1971b)	80–400
Upper Volta	Roose (in Josens, 1983)	60
Australia	Williams (1968)	200–300

between 250 and 260,000 t km^{-2} a^{-1}, with most studies reporting values between 2000 and 20,000 t km^{-2} a^{-1}.

Even when one combines the amount of soil translocation involved in the formation of both termite mounds and sheetings the rates still tend to be relatively low in comparison with worm cast formation. Josens (1983) reports combined rates of 180 and 120 t km^{-2} a^{-1} from Senegal and Upper Volta respectively.

By continuing studies of rates of termite-caused soil translocation and estimates of rates of land surface denudation it is possible to estimate the speed at which termites can build a new soil layer (table 4.15). There is a great variability in the proposed rates from different areas, and on this basis a 1 m thick soil layer might take anything between 1000 and 40,000 years to form.

The role of other organisms in sediment transport

Although termite and worms may be among the most widespread and effective of the organisms that translocate sediment, it is undoubtedly true that there are many instances where other types of organisms also play a major role. Of particular significance is the role of burrowing animals, such as gophers, moles, mole rats and the like. They can excavate and mound up appreciable amounts of material, and the material may then be washed downslope by rainsplash, surface runoff and other processes.

It is, however, crucial that estimates be made of the relative importance of translocation material made by organisms and that achieved by other inorganic processes. Several recent studies have attempted to do this for some medium-sized mammals, and they have tended to show that the role of such mammals is relatively quite modest. In forested areas in the Belgian Ardennes, for example, Voslamber and Veen (1985) calculated that the mass transport effected directly by badgers was 'insignificant', and that their production of 'available sediment' was 'relatively unimportant' in this temperate humid forest environment. Likewise, in a non-forested area of Central California, Black and Montgomery (1991) recorded the burrowing activity of gophers, and compared this with long-term sediment transport rates determined by the study of colluvium volumes of known age. The amount of sediment transport they achieved was nearly an order of magnitude less than the average long-term rates.

Plainly, however, it is extremely difficult to generalize about this issue given our present state of knowledge. The amount of material moved by mammals will vary according to a multitude of factors, which include the nature of the substratum into which they burrow, the density of their populations, the species of the burrowing animals, and the efficiency with which other processes operate.

Another potentially important type of organic contribution to sediment movement is that caused by the growth, collapse and death of plants. For example, sediment movement on slopes is achieved by the process of tree fall. As trees topple over, their root mat and associated rock debris is dislodged and a portion of it moves downslope. Methods for trying to estimate the quantities of material involved are described by Denny (1956) and Mills (1984). Self evidently, rates will depend on such factors as the frequency of tree fall, the size of the root mat, the density of trees, and the nature of the slope and the material of which it is composed. There have been far too few studies of this phenomenon to make any serious generalizations about either rates or the factors that control their operation. However, the two studies for which there are some quantitative data both show that it is a moderately important cause of sediment movement downslope, though the rates differ by an order of magnitude! The rate given by Denny is 125 mm 1000 a^{-1} of surface erosion, compared with a rate of 13 mm 1000 a^{-1} given by Mills.

Peatbog growth

Peatlands, formed by the accumulation of organic matter, are extensive and important geomorphological features. They may cover something like 3.4 per cent of the Earth's land surface, contain something like 240 Gt of carbon, and are major repositories of detailed information about environmental changes.

In geological terms, peat formation is a speedy process. Several estimates have been presented of the sorts of rates that occur on average. Moore and Bellamy (1974, p. 105) suggest a value of between 200 and 800 mm 1000 a^{-1}, while Clymo (1991) gives a value of 500 to 1000 mm 1000 a^{-1}, Cameron (1970) quotes orders of bog formation for North America of 1000 to 2000 mm 1000 a^{-1}. If these values are taken as representative then it is likely that deposits of peat some 2 to 20 m thick could accumulate in the course of the Holocene.

The rates quoted above are estimated by obtaining dates for specific layers in the stratigraphy of peat profiles and thus calculating rates of peat build up. However, given the environmental

changes that have occurred during the Holocene, it is probable
that the actual rates of formation have varied through time. Among
the controls of rate are the speed at which organic matter accumulates ('litter fall'), the speed at which it decays, the speed at
which compaction occurs, the speed at which erosion operates,
and the speed at which nutrients are added. These controls will
vary according to such factors as rainfall, evapotranspiration,
subsoil permeability, temperature, human activity, etc. Walker
(1970), by collecting data from a variety of sources, attempted to
show the temporal changes that had taken place in the Holocene
rate of peat formation at a variety of British and Irish sites, and
this has now become an issue of some debate as scientists argue
about the extent to which peatbog formation is the result of
human interference (see, Bell and Walker, 1992, for a discussion).
Extreme care needs to be taken in assessing rates of accretion
during the Holocene as they can be greatly influenced by auto-
compaction processes (Aaby and Tauben, 1974). Indeed, Svensson
(1988) believed that this process helped to explain the apparently
accelerating rate of accretion found at a dated site in southern
Sweden, where the mean growth rate for the last 5000 years has
been 840 mm 1000 a^{-1}, but for the last 1000 of these years has
been 1550 to 2320 mm 1000 a^{-1}.

Nonetheless, some rates of change in peat accretion are undoubtedly 'real', and probably reflect the influence of changes in
climate during the course of the Holocene. A particularly interesting example of this comes from Canada, where Ovenden (1990)
has analysed both the spatial and temporal patterns of peat accumulation. Rates of accumulation tend to be higher in the boreal
wetlands of temperate Canada than they are in the subarctic regions. The southern boreal and maritime regions of Canada are
more conducive to peat formation on suitable sites than high boreal
and subarctic regions. However, in the northern regions peat
accumulation rates were higher in the mid-Holocene than they
are now, being at a maximum between 6000 and 2500 yr BP. Peat
accumulation is low at many of these sites now and declined
rapidly at many of them from 3250 to 2000 yr BP. Ovenden associates the phase of mid-Holocene peat growth to a favourable
climatic regime, when temperatures were slightly depressed and
annual precipitation increased. This reduced soil moisture deficits
and raised regional water tables about 6000 to 3000 yr BP. Figure
4.7 tabulates the timing of rapid peat growth at sites near the
treeline in the Northwest Territories.

Conclusion

Over the last three decades slopes have taken their rightful place
in Geomorphology and many techniques have been developed to
study the rate of operation of slope processes. In spite of this

Figure 4.7
*Frequency of sites near
the treeline in the
Northwest Territories of
Canada where peat
accumulation exceeded 0.2
mm per year during each
250-year interval of the
Holocene. The shaded
portion indicates
accumulation at sites east
of 120°W (source:
Ovenden, 1990, fig. 3).*

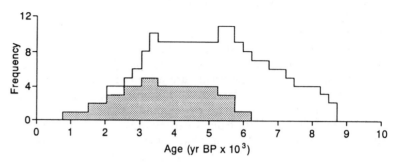

expansion of effort, we are still far from being able to make any
very firm statements about issues such as the speed at which different mass movement processes tend to operate under different
climatic conditions. Indeed, given the complexity of factors that
control rates of process operation on slopes, and given the way in
which their relative importance varies at different spatial scales, it
may well prove to be an intractable problem.

One thing that is clear, however, is that hill-slope surfaces are
particularly prone to human impacts through land cover change
(e.g. deforestation, agriculture, urbanization, fire, etc). Equally,
other organisms, most notably worms and termites, play a major
role in sediment movement.

FIVE

The power of wind

Wind erosion: introduction

As recently as two decades ago, the erosive action of wind in the desert environment was accorded a limited role, and desert landscapes were seen to be largely inactive under present climatic regimes (see Goudie, 1989, for a review).

Such a view had not always been the norm, for at the turn of the century there was a phase of what has been termed extravagant aeolation (Cooke and Warren, 1973). This had its roots in the work undertaken in Africa by French and German geomorphologists (e.g. Walther, 1900; Passarge, 1904), but was put forward in its most exuberant form in the USA by Keyes (1913), who believed that material weakened by thermoclasty (insolation weathering) would be evacuated by wind and deposited as dust sheets on desert margins. He argued that the end result of such activity would be the formation of great plains, mountain ranges without foothills, and towering eminences.

The reasons for the decline of aeolianist views were many. Firstly, the great pediment landscapes of the American deserts were seen, following the work of McGee (1897) and others, as being attributable to planation by sheetflood activity. Secondly, many desert landscapes were thought to have been moulded by fluvial processes that had been more powerful and widespread during the pluvial phases that were held to be a feature of the Pleistocene. Current arid conditions were widely thought to be a relatively recent and short-lived phenomenon caused by post-glacial progressive desiccation (Goudie, 1972). Thirdly, doubt was expressed about the power of thermoclasty as a process capable of preparing desert surfaces for subsequent aeolian attack. Such doubt arose because of the experimental work of Blackwelder (1933) and Griggs (1936). Fourthly, it was widely held that lag gravels (stone pavements) and salt and clay crusts would limit the extent to which aeolian processes could cause lowering of surfaces, and that it was impossible for aeolian processes to cause excavation of surfaces

Table 5.1   Methods for determining rates of aeolian excavation and abrasion

1. Relief inversion in areas of suspendritic drainage
2. Deflation of lake beds of known age
3. Exposure of bodies, etc. formerly buried at a known depth
4. Monitoring individual events with stakes, etc.
5. Examination of aeolian sediment loadings in deep-sea cores
6. Use of artefacts (bricks, leucite rods, etc.) to determine abrasion
7. Micro-erosion meter (MEM)

below the water-table (Cooke and Warren, 1973, p. 251). Fifthly, it became apparent that many of the world's great dust sheets, in North America, China, and the former USSR, were the product of deflation from glacial areas rather than from deserts. Glacial grinding was thought to be the most efficient way of producing silt-sized quartz particles (Smalley and Krinsley, 1978). Sixthly, it was recognized that not all deserts had adequate supplies either of abrasive sand or of frequent high-velocity winds for wind erosion to be achieved with any degree of facility. Finally, features that were conceded to have an aeolian origin (e.g. yardangs, ventifacts and pedestal rocks) were thought to be minor, bizarre embellishments of otherwise fluvial environments, whilst other possibly aeolian features (notably stone pavements and closed depressions) were also explicable by other means. Stone pavements, for example, could be the product of the removal of fine sediments by sheetflood activity or they could result from vertical sorting processes associated with wetting and drying, salt hydration or freezing and thawing (Cooke, 1970). Deflational removal of fines to leave a lag was just one possible formative mechanism. In the same way, closed depressions could be attributed to wind excavation, but might also be explained by tectonic, solutional or zoogenic processes (Goudie and Thomas, 1985).

Many of these objections to widespread and effective aeolation are fundamentally cogent and sound, but nonetheless in the last two decades the power of wind to erode desert landscapes has been reassessed. It has been shown to have a considerable significance in moulding rock surfaces to produce yardangs, in causing relief inversion, in excavating depressions, and in causing dust storms and deflation. Why has there been this resurgence in ideas on aeolation? Two of the major stimuli to such a reassessment have been the search for analogues of Martian features, and the increasing availability of remote sensing images from aircraft and from space vehicles. Also important, however, have been some field studies of aeolian abrasion rates, and the study of dust storms and deflation by means of analysis of meteorological data, satellite images and sediment cores.

Estimation of rates of aeolian denudation requires the development of some particular techniques (table 5.1). In areas where wind excavation has caused the development of inverted relief (suspendritic drainage) the dating of the upstanding channel deposits and the surveying of the degree of inversion may permit the estimation of rates over periods of some thousands of years. Similarly, if lake bed remnants of known age give a clear picture of the former height of a lake floor, then it is possible to calculate the rate of lake floor deflation. Alternatively, dated archaeological remains, including buried corpses, may become exposed by wind excavation. The presence of aeolian dust-derived sediments in ocean cores may enable the reconstruction of long-term deflation histories from the neighbouring desert source area. Over shorter time spans, rates of surface lowering can be directly monitored with erosion pins, stakes, micro-erosion meters and buried base-plates (Sarre, 1989). Short-term rates of aeolian abrasion can be estimated through the degree of furrowing developed on bricks and other materials of known age.

The controls on wind erosion on agricultural land are often expressed as a Wind Erosion Equation

$$E = f\ (C,I,L,K,V)$$

where E is the potential erosion loss, C is a local climatic index, I is a soil erodibility index, L is a factor relating to field shape in the prevailing wind direction, K is a ridge roughness factor for ploughed ground, and V is a vegetation cover index. The equation was developed initially for the Midwest of the USA (see Woodruff and Siddoway, 1965) and drew attention to the factors which could be manipulated by farmers (namely, I, L, K and V).

The climatic factor (C) was a simple combination of two key climatic variables: annual wind speed and a moisture index. Plainly, dry windy areas are likely to be most susceptible to wind erosion. The soil erodibility factor (I) is more complex, and needs to be seen in terms of both individual grain size characteristics and aggregate characteristics. Fine sands and silts are likely to be most susceptible, partly because of the relatively low velocities required for their entrainment, but also because clay's presence tends to produce wind-stable clods. The presence of large clods reduces the risk of wind erosion. The fetch distance over which the wind acts (L) is related to field size and the presence or absence of shelter belts of differing heights, spacing and permeability. The ridge roughness factor (K) is based on the experimental observation that the rougher the surface, up to about 6 cm, the lower the windspeed at the surface. Thus furrows at right angles to the wind will tend to dampen down rates of wind erosion. The vegetation factor (V) is absolutely fundamental, for a dense vegetation cover, especially if like grass it has short stalks and narrow leaves, does more than anything else to reduce erosion rates.

Yardang and ventifact formation

A landform type whose morphology suggests that it is primarily moulded by wind erosion is the yardang (plate 5.1).

Plate 5.1
A yardang developed in unconsolidated Pleistocene sediments near Dushanbe in the Tajik Republic. Very little monitoring of rates of yardang formation in particular and of wind erosion in general has as yet been undertaken.

Small yardangs appear to have the ability to form quickly in Holocene lake deposits. Many examples from the Sahara post-date the so-called Neolithic pluvial, the last time that many of the lake basins filled with water. This suggests that rates of incision have been between 400 and 4000 mm 1000 a^{-1}. The Lop Nor yardangs from high Asia may have been eroded since the fourth century AD, indicating a rate of erosion as high as 20,000 mm 1000 a^{-1} (McCauley et al., 1977). The yardangs from the Californian playas have been eroded at a maximum rate of as much as 23,000 mm 1000 a^{-1}. It needs to be stressed, however, that these data apply to highly erodible and largely unconsolidated and unlithified lacustrine fill materials, and such rates should not uncritically be applied to yardangs grooved into bedrock. Some of the great rock yardangs of the Sahara may pre-date the Pleistocene.

Some more general rates of lake basin deflation are available from the Sahara. Boyé et al. suggested that the Sebkha Mellala had been deflated during the late Holocene at a rate of about 410 mm 1000 a^{-1}, while Riser (1985), working on the Araouane Basin of Mali, where kitchen middens and skeletons had been exposed, found a rate of ablation of 60 cm over the last 6500 years (equivalent to c 92 mm 1000 a^{-1}).

Although wind abraded landforms ranging in scale from small ventifacts to huge yardangs have long been studied, there are surprisingly few pieces of information on the rate at which wind can abrade rocks and related materials. Sharp (1949) described ventifacts of granite, gneiss and quartzitic sandstone from the Big Horn Mountains of Wyoming and inferred a rate of about 1000 mm 1000 a^{-1}. Hickox (1959) studied ventifacts in the Annapolis Valley of Nova Scotia (Canada) which were composed of vein quartz, quartzite and indurated sandstone. He considered that the faceting of the ventifacts had taken place in less than 10 years from exposure by quarrying, giving a rate of 10,000 mm 1000 a^{-1}. McCauley et al. (1979) discovered sandstone abraded hearth stones of Upper Palaeolithic age in Egypt which were abraded to a depth of 2 to 4 cm, giving a rate of about 1 to 2 mm 1000 a^{-1}. From examination of a wind-abraded dacite boulder field at Mono Craters, California, Williams (1981) obtained a rate of abrasion of 1000 mm 1000 a^{-1}. Finally, Sharp (1964) measured abrasion rates for bricks and other materials at Garnet Hill, California, and found rates of around 50,000 mm 1000 a^{-1}. The present author has seen buildings made of brick in Kolmanskop, Namibia, cut through by wind abrasion at their bases. If the buildings have been cut through since they were abandoned in the early 1950s the rate of cutting is about 225 mm in 40 years, giving a long-term rate of over 5000 mm 1000 a^{-1}. If the cutting through is regarded as having taken place since construction in the 1900s, then the rate is of the order of 2500 mm 1000 a^{-1}.

Plainly, there is a great range in these values. However, it is also plain that in suitable environments (dry, windy and near to the ground) materials of the strength of brick can abrade really very rapidly under 'normal' desert conditions. There are very few data on rates of abrasion in polar desert environments, though there are many qualitative descriptions of its effects. Micro-erosion meter studies in the Larsemann and Vestfold Hills of Antarctica, where wind erosion and salt wedging operate in conjunction, have, however, revealed surface lowering rates of 15 to 20 mm 1000 a^{-1} on dolerites and gneisses (Spate et al., in press).

Current rates of dust deposition on land

Wind deposition rates are rather easier to determine than are wind erosion rates. A large amount of data has accumulated in recent years but considerable care needs to be exercised in the use of data on current rates of dust deposition on land. The rates are often based on very short periods of observation, there may be a tendency either to miss extreme events or to report then when they have been successfully monitored because of the spectacular nature of

Table 5.2 Rates of dust deposition on land (current)

Source	Location	Rate t km⁻² a⁻¹	**mm 1000 a⁻¹
Mediterranean region			
Loÿe-Pilot et al. (1986)	Corsica	14	16
Yaalon and Ganor (1975)	Israel	22–83	25–93
Bucher and Lucas (1984)	Pyrenees	18–23	20–26
USA			
Smith et al. (1970)	High plains	65–85	73–96
Brown et al. (1968)	Kansas	6.6–8.6	7–10
Van Heuklon (1977)	Illinois	100	112
Péwé et al. (1981)	Arizona	54	61
Gile and Grossman (1979)	New Mexico	9.3–125.8	10–141
Muhs (1983)	California	24–31	27–35
Miscellaneous			
Maley (1980)	Chad	109	122
McTainsh and Walker (1982)	N. Nigeria	137–181	154–203
Safar (1985)	Kuwait	100	112
Tiller et al. (1987)	SE Australia	5–10	6–11

** Calculated on bulk density of dust of 0.89 g cm⁻³.

the results, and the data are generated by many different techniques.

Nonetheless, as table 5.2 shows, there are some data available which give a broad indication of the effectiveness of aeolian dust deposition at the present day. The range of values is very roughly between 6 and 200 t km⁻² a⁻¹. It is clear from these data that areas like the Mediterranean Basin, the High Plains of the USA and the borders of the Sahara, all receive appreciable aeolian additions at the present time, and that these are probably of the same order of magnitude as fluvial outputs in many cases. Almost certainly such rates of deposition indicate that aeolian dust is an important contributor to the speed and nature of soil and stone pavement formation.

The amount of dust material moved and deposited by individual events may also be appreciable (Goudie, 1978, table 6, p. 300): the 1901 dust fall over North Africa, for example, is estimated to have deposited 15×10^7 tons and the 1903 dust fall over England is estimated to have deposited ~10^7 tons of sediment. When expressed in tons of sediment deposited per square kilometre, rates can reach as high as 500 tons km⁻² (Idaho), 455 tons km⁻² (North Africa), 300 tons km⁻² (Nebraska), 162 tons km⁻² (Colorado) and 126 tons km⁻² (Caspian) on desert margins. Quantities fall off rapidly towards more humid areas. Data for a major dust storm in the USA and Canada in 1933, for instance, show deposition rates of 39 tons km⁻² in Kansas, 13.5 tons km⁻² in east Nebraska, but only 0.58 tons km⁻² in New Hampshire. Nonetheless, moderately high dust falls have been recorded in Europe and Britain (3.83–195 tons km⁻²).

It is extremely difficult to estimate the quantities of dust produced

Table 5.3 Estimates of dust production

Location	Source	Annual quantity (millions of tons)
World	Schütz (1980)	up to 5000
World	Petersen and Junge (1971)	500
World	Joseph et al. (1973)	128 ± 64
Sahara	Schütz (1980)	300
Sahara	d'Almeida (1986)	630–710
Aral Sea area	Grigoryev and Kondratyev (1980)	75
Southern Mediterranean	Yaalon and Ganor (1975)	20–30
Sourth-western USA	Gillette (1980)	4
Mediterranean	Joseph et al. (1973)	3.2 ± 1.6

in different areas, and to come up with a global picture of dust production (see Prospero, 1981, for a discussion). Estimates of global aeolian dust contributions to the atmosphere vary widely, but may reach 5000 m tons per year (Schutz, 1980). The Sahara contributes the largest proportion of this total, and as table 5.3 shows, other deserts are relatively much less important.

Dust into the oceans

In comparison with the situation on land, there are relatively few observational data on long-term dust accumulation rates over the oceans. There are figures for the Pacific Ocean (Pye, 1987, pp. 90–1) and they are consistently low in comparison with rates determined on land. Given the distance decay in rates as one moves away from sources this is hardly surprising. Rates are of the order of 0.1 to 1.5 t km^{-2} a^{-1}.

However, it is also possible to obtain a long-term measure of dust additions to the oceans by undertaking studies of the sedimentology of deep-sea cores. The same data can also be used to compare dust accumulation rates during glacials and interglacials.

The difference in accumulation rates through time is instructive. In particular it is possible to identify differences in the amount of dust being deposited in pre-Pleistocene times in comparison with that during the Pleistocene, and it is also possible to draw contrasts between the glacial and interglacial phases of the Quaternary.

With regard to the differences between the glacials and interglacials, the results largely confirm the information obtained from loess sections and ice cores to the effect that the world was extremely dusty during cold phases. For example, on the basis of their analysis of cores from the Arabian Sea, Sirocko et al. (1991) suggested that dust additions were around 60 per cent higher in glacials than in post-glacial time. Likewise, also working in the Arabian Sea, Clemens and Prell (1990) found a positive correlation between global ice volume (as indicated by the marine $0^{18}0^{16}$

record) and the accumulation rate and sediment size of dust material. Kolla and Biscaye (1977) confirmed this picture for a larger area of the Indian Ocean, and indicated that large dust inputs came off Arabia and Australia during the last glacial. In the Atlantic Ocean offshore from the Sahara, at around 18,000 years ago the amount of dust transported into the Ocean was augmented by a factor of 2.5 (Tetzlaff et al., 1989, p. 198).

At a longer time-scale there is some evidence that dust activity increased as climate deteriorated during the late Tertiary. In the Atlantic off West Africa, Pokras (1989) found clear evidence for increased terrigenous lithogenic input at 2.3 to 2.5 million years ago while Schramm (1989) found that the largest increases in mass accumulation rates in the North Pacific occurred between 2 and 3 million years ago. This coincides broadly with the initiation of Northern Hemisphere glaciation. However, no such link has been identified in the southern Pacific Ocean (Rea, 1989). The lengthiest analysis of dust deposition in the oceans has been undertaken by Leinen and Heath (1981) on the sediments of the central part of the North Pacific. They have demonstrated that there were low rates of dust deposition 50–25 million years ago. This they believe reflects the temperate, humid environment that was seemingly characteristic of the early Tertiary and the lack of vigorous atmospheric circulation at that time. From 25 to 7 million years ago the rate of aeolian accumulation on the ocean floor increased, but it became greatly accelerated from 7 to 3 million years ago. However, although there is thus an indication that aeolian processes were becoming increasingly important as the Tertiary progressed, it was around 2.5 million years ago that there occurred the most dramatic increase in aeolian sedimentation. This accompanied the onset of northern hemisphere glaciation, which has been a feature of the last two to three million years. There thus appears to be evidence that the Late Cainozoic ice age was a time when dust storms became more frequent than they had been for the previous 70 million years.

Dust deposition as recorded in ice cores

A third major source of long-term information on rates of dust accretion is the record preserved in long-ice cores retrieved either from the polar ice-caps or from high-altitude ice domes at lower altitudes.

Because they are generally far removed from source areas, the actual rates of accumulation are generally low, but studies of variations in micro-particle concentrations with depth do provide insights into the relative dust loadings of the atmosphere in the last glacial and during the course of the Holocene. Thompson and Mosley-Thompson (1981) drew together a lot of the material that was published at the time they wrote, and pointed to the great differences in micro-particle concentrations between the Late Glacial

Plate 5.2
Loessic profiles demonstrate the rapidity with which aeolian silts accumulated during the Pleistocene glacials. Palaeosols within the profiles indicate periods of relative stability during interglacials. This example is from the Tajik Republic.

and the Post-glacial. The ratio at Dome C Ice Core (E. Antarctica) was 6:1, for the Byrd Station (W. Antarctica) 3:1, and for Camp Century (Greenland) 12:1. Briat et al. (1982) maintained that at Dome C there was an increase in micro-particle concentrations by a factor of 10 to 20 during the last glacial stage, and they explain this by a large input of continental dust. The Dunde Ice Core from High Asia (Thompson et al., 1990) also shows very high dust loadings in the Late Glacial and a very sudden fall off at the transition to the Holocene.

Loess accumulation rates

By measuring and dating loess sections (plate 5.2) it has been possible to estimate the rate at which loess accumulated on land during the Quaternary (see table 5.4). Loess can be dated by a range of techniques (including thermoluminescence, palaeomagnetism, and radiocarbon) and many impressive sections are available for study. The presented data may somewhat underestimate total dust fluxes into an area because even at times of rapid loess accumulation there would have been concurrent losses of material

Table 5.4 Loess accumulation rates for the Late Pleistocene

Location	Rate (mm 1000 a^{-1})
Negev (Israel)	70–150
Mississippi Valley (USA)	100–4000
Uzbekistan	50–450
Tajikistan	60–290
Lanzhou (China)	250–260
Luochuan (China)	50–70
Czechoslovakia	90
Austria	22
Poland	750
New Zealand	2000

Source: from various sources in Pye (1987), and Gerson and Amit (1987).

as a result of fluvial and mass movement processes. Solution and compaction may also have occurred.

The data in table 5.4 show a range of values between 22 and 4000 mm 1000 a^{-1}, but Pye (1987, p. 265) believes that at the maximum of the last glaciation (at c.18,000 years ago) loess was probably accumulating at a rate of between 500 and 3000 mm 1000 a^{-1}, and suggests that 'Dust-blowing on this scale was possibly unparalleled in previous Earth History.' By contrast, he suggests that 'During the Holocene, dust deposition rates in most parts of the world have been too low for significant thicknesses of loess to accumulate, although aeolian additions to soils and ocean sediments have been significant.' Pye also hypothesizes that rates of loess accumulation showed a tendency to increase during the course of the Quaternary. Average loess accumulation rates in China, Central Asia and Europe were of the order of 20–60 mm 1000 a^{-1} during Matuyama time, and of the order of 90–260 mm 1000 a^{-1} during the Brunhes epoch. He also points out that these long-term average rates disguise the fact that rates of loess deposition were one to two orders of magnitude higher during Pleistocene cold phases, and one or two orders of magnitude lower during the warmer interglacial phases when pedogenesis predominated.

The high rates of dust deposition in glacial times over China are also reflected in the rate of dust accumulation over Japan, a country which receives much dust from China, Mongolia and Korea. Studies by Inoue and Naruse (1987) indicate that the flux of dust to the Japanese land surface in modern times averages 3.6 to 7.1 mm 1000 a^{-1}, in comparison with 13.5 to 22.9 mm 1000 a^{-1} in the last glacial, a more than three-fold decline.

The rate of advance of dunes

Having discussed rates of wind erosion and dust deposition, it is appropriate to consider the rates at which another aeolian phenomenon – the dune – moves.

Table 5.5 Rates of barchan migration

Location	Source	Height (m)	Rate of advance (m a⁻¹)
Namib	Kaiser (1926)	33	48
	Endrody-Younga (1982)	8–10	43
	Barnard (1973)	11–33	8.4
	Ward (1984)	10–20	0.8–6.4
	Slattery (1990)	8.25	14.6
Middle East	Fryberger et al. (1984)	30+	15
	Tsoar (1974)	3.5	6.3
	Embabi (1982)	5.52	48.35
	Shehata et al. (1992)	7–11	8.7–11.3
The Americas	Finkel (1959)	3.7	15.1
	Hastenrath (1967)	3–4	14.2–30.8
	Long and Sharp (1964)	6.1	18.7
	Norris (1966)	11	13.4
	Smith (1970)	6	20

There are more data for rates of dune migration with respect to barchan dunes than for any other dune type. This is scarcely surprising given that such dunes are readily identifiable, can be relatively precisely defined with respect to their shape, and do migrate at rates which are sufficiently large to be measurable over a finite period of time.

Data have previously been assembled and analysed by Lancaster (1989) and Thomas (1992) and these form the basis of table 5.5. It becomes apparent from this information that average annual rates range over several orders of magnitude from the 0.08–0.11 m a⁻¹ for a crescentic draa (dune plinth) in the Algodones erg to 80 m a⁻¹ for a small Peruvian barchan. Among the factors that may control these differences are surface characteristics (e.g moisture, vegetation cover, and sand supply), wind strength and direction, and dune size. The last of these factors is generally accorded the greatest significance, for as Thomas (1992) put it, 'Size dependency is not surprising as the whole dune migrates forward by a "rolling" process whereby sediment exposed at the foot of the stoss side is transported up to the dunecrest to be deposited on the slipface. . . . The larger the dune, therefore, the more sediment that has to be moved per unit of forward movement.' The rate of movement can therefore be expressed as the following formula (after Bagnold, 1941):

$$C_r = (q_c - q_t)/h_{\gamma_p}$$

where C_r is the advance rate, q_c and q_t are respectively the mass transport rates at the dune crest and the dune base, h is the dune height and γ_p is the bulk density of the sediment.

Figure 5.1 shows the relationship between dune height and rate of migration as demonstrated by Slattery (1990) in the Namib

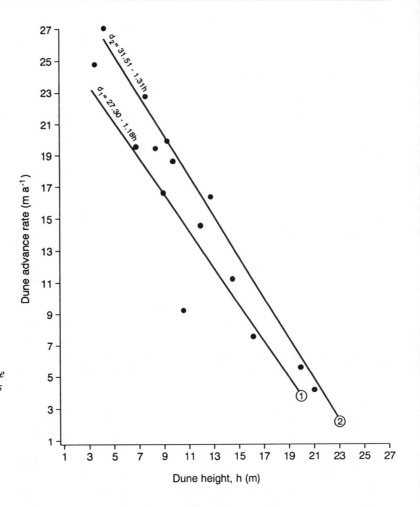

Figure 5.1
Relationship between dune advance rate (d) in metres per year for barchan dunes in the central Namib Desert, Namibia, between 1961 and 1976 (line 1) and for 1976 to 1988 (line 2) (source: Slattery, 1990, fig. 6).

(plate 5.3). The relationship shown is a clear linear one. However, there has been some debate on the form of the relationship in the literature (see Cooke, Warren and Goudie, 1993, pp. 339–40) and doubts have been expressed as to whether it applies at either end of the size range. The rate of movement of very small barchans (i.e less than 1 m high) may be due to their higher bulk densities or because of the more variable wind environment for smaller dunes, sheltered by large ones. At the other end of the size range, as figure 5.2 demonstrates, in many cases the relationship should be exponential rather than linear. Larger dunes seem to reach a plateau in their rate of movement beyond which size seems to make very little difference. This may perhaps be due to greater wind speed-up on higher dunes.

This is not to say, however, that the other factors are not significant in determining the rate of advance. For example, Lancaster (1989, p. 94) has commented on the very high rates of barchan

Plate 5.3
The fastest-moving dune form is the individual barchan, seen here encroaching on an abandoned railway line near Swakopmund in Namibia.

migration in the Namib for dunes of any particular size in comparison with many other areas. He attributes this to the extremely vigorous wind energy conditions that exist in the coastal environment of the area between Luderitz and Alexander Bay.

Equally, in his long-term study of the rates of barchan dune movement in the Western Desert of Egypt, Embabi (1986/7) found that dunes in the Kharga Depression had between 1930 and 1961 moved on average some 280 m, while those at Dakhla had averaged some 170 m. The explanation he offered was in terms of differences in sand drift potential between the two areas with the Kharga figure being 245 Vector Units compared with 73 Vector Units at Dakhla. The meaning of these units is discussed in the next section.

Data on the movement of linear dunes are much more sparse and much more difficult to interpret. One of the problems is that linear dunes may move in one of three ways: dune extension at the leeward end, along dune displacement of crestal 'peaks', and lateral displacement of the dune under the influence of one of the oblique formative wind components (Thomas, 1992, p. 34). In table 5.6 data are presented for two types of linear dune: seifs and

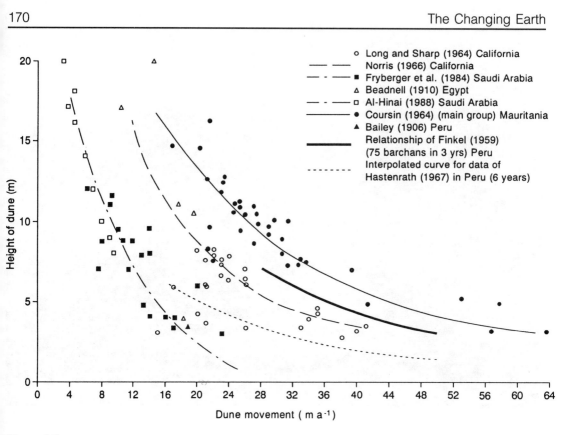

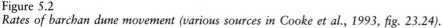

Figure 5.2
Rates of barchan dune movement (various sources in Cooke et al., 1993, fig. 23.24).

linear ridges. Thomas (1992, p. 35) believes that if these data are representative of seif and linear ridge dynamics as a whole they have considerable implications for the nature of linear dune activity:

First, linear dunes appear to undergo little movement of the dune form as a whole. Extension at downwind termini affects only a small part of the total dune body, given that linear dunes are commonly several kilometres long and that it is not uncommon for them to extend for tens of kilometres (Thomas, 1989). Second, empirical evidence indicates that even the movement of relatively active seif dunes is restricted essentially to the crestal zone. Thus expressions of this movement occur within the dune and as both Tsoar (1978) and Livingstone (1989) have shown, the plinths and lower slopes of the dunes remain stable. Even when Rubin's (1990) findings are taken into account, linear dunes can be regarded as essentially stationary features, though they are by no means inactive, and the role of lateral movement both of linear ridges (Rubin, 1990) and even possibly seif dunes (Tseo, 1990) warrants further investigation.

Coastal dunes seem to be capable of relatively rapid inland movement, as is made evident by the records of buildings being catastrophically overwhelmed, though much depends on the nature of the coastal vegetation and the extent to which it has been

Table 5.6 Dune activity rates, linear forms

Location	Dimensions (m)		Movement (m a⁻¹)			Source
	Height	Width	Along-dune	Elongation	Lateral migration	
Seif dunes						
Sinai	12–14	50	8.4	14.5	Crest 5–7* Plinth 0	Tsoar (1978)
Namib	50	350			Crest 15* Plinth 0	Livingstone (1989)
Namib				>1.85		Ward (1984)
Linear Ridges						
Strezlecki, Australia	10–20				50–100 m since Pleistocene	Rubin (1990)
William River, Canada **	12–30	120–160			0.2–0.6	Carson and MacLean (1986)

* Values are seasonal reversals, resulting in no net change.
** Dunes termed 'hybrid dunes' by authors, but regarded as linear ridges by Rubin (1990).
Source: from Thomas (1992) table 2.

denuded by human activities. Kukal (1990) suggests that because of the high energy conditions on exposed coastlines coastal dunes will tend to migrate quickly, and he gives figures for The Landes of France, the Kursk Peninsula and for Denmark that range from 3 to 25 m a⁻¹.

Sand sea growth and sand movement through ergs

Palaeoclimatic and stratigraphic studies have shown that many of the world's sand seas are of considerable antiquity, and many of them may date back to the late or mid Tertiary. There is some evidence, according to Cooke et al. (1993), that older sand seas, like the Namib, which may date back to the Miocene, contain more sand than younger ones like those in Australia. It is possible to use measurements of sand or dune movement and of the volumes of sand in a dune field or sand sea (erg) to arrive at estimates of its age. The following illustrations come from Cooke et al. (1993, p. 409). Ball (1927) calculated that if dunes in the Abu Moharik dune field in Egypt moved at 15 m a⁻¹ the dune belt itself would have been 35,000 years old. Using the same estimate for dune movement (15 m a⁻¹), and the known distance of advance from the coastal sand source, Warren (1988) calculated that the Holocene dune field in the Wahiba Sands was about 8000 years old. In coastal Baja California, Inman et al. (1966) estimated the advance of 6 m-high dunes to be 18 m a^{-1}, which gave an age for the 12 km-wide dune field of about 18,000 years. Sweet et al. (1988) calculated the migration rate for the Algodones dune field at .0135 m a^{-1}. They maintained that the dune field migrated more slowly than its constituent dunes. Other calculations of dune field ages have been made by Shinn (1973) and Sarnthein and Walger (1974).

Sand seas are inevitably older than dune fields. Using estimates of dune bulk transport Fryberger and Ahlbrandt (1979) calculated that a sand sea of 10^4 km^2 area × 100 m depth would need between 13.5 million and 4.17 million years to grow.

Rates of bulk transport may not be the most appropriate for this purpose, because sand seas grow as much by the accretion of sand to existing bedforms as by the inward migration of new ones. Using estimates of sand, rather than dune movement, Wilson (1971) estimated the age of the Great Eastern Sand Sea in Algeria to be 1,350,000 years.

An additional temporal requirement for the growth of large bodies of aeolian sand is continued or repeated aridity. Continued aridity appears to be one of the reasons for the growth of the large Namib Sand Sea (Lancaster, 1989). In the northern Sahara, on the other hand, the great volume of sand in the Great Sand Seas is more a function of the repeated alternation of phases of humidity and water erosion (to supply sand), and aridity and aeolian activity (for aeolian winnowing and spatial concentration).

The rate at which sand moves through any particular erg (sand sea) is related to the area's wind energy. The amount of sand migration (or sand drift) can be calculated by using a modified Lettau equation (Fryberger, 1979, p. 141):

$$Q \propto V^2 (V - V_t) \cdot t$$

where Q is a proportionate amount of sand drift, V is average wind velocity at 10 m height, V$_t$ is the impact threshold wind velocity, and t is the time that the wind blew, expressed as a percentage in a wind summary.

Values derived from the above formula by substitution are termed drift potentials (DP), which are numerically expressed in vector units (VU). The directions and magnitude of the vector resultants of drift potentials from the different compass directions (normally 16) are known as the resultant drift direction (RDD) and the resultant drift potential (RDP) respectively. It needs to be stressed that these are all related to *potential* sand drift, and that other factors may control *actual* sand drift (e.g. surface conditions and the nature of the sand involved in the transport process).

DP values enable a comparison to be made between different desert areas. The values show a six-fold range, and Fryberger divided ergs into three classes (table 5.7) with the high-energy wind environments having values in excess of 400 vector units

Table 5.7 Average monthly and annual drift potentials for thirteen desert regions, based on data from selected stations

Desert region	Number of stations	Jan.	Feb.	Mar.	Apr.	May	June	July	Aug.	Sept.	Oct.	Nov.	Dec.	Annual drift potential
High-energy wind environments														
Saudi Arabia and Kuwait (An Nafūd, north)	10	35	39	52	54	51	66	49	33	20	18	16	25	489
Libya (central, west)*	7	40	42	48	64	51	41	20	18	24	24	22	37	431
Intermediate-energy wind environments														
Australia (Simpson, south)	1	43	40	27	17	13	10	18	26	52	56	46	43	391
Mauritania	10	45	49	45	38	33	40	26	19	20	20	19	30	384
U.S.S.R. (Peski Karakumy, Peski Kyzylkum)	15	39	41	43	43	33	25	22	21	23	23	24	29	366
Algeria	21	21	27	37	48	32	27	18	13	15	16	16	23	293
South-West Africa (Namib)	5	8	2	6	17	13	50	19	22	27	44	17	12	237
Saudi Arabia (Rub' al Khali, north)	1	23	28	53	32	20	30	. . .	. . .	. . .	1	'7	7	201
Low-energy wind environments														
South-West Africa (Kalahari)	7	14	11	8	10	9	11	18	24	26	26	17	18	191
Mali (Sahel, Niger River)	8	9	12	14	12	19	22	15	9	10	5	5	7	139
China (Gobi)*	5	9	11	16	23	20	11	7	5	5	5	7	8	127
India (Rājasthān, Thar)*	7	2	2	5	5	10	21	19	9	5	2	1	1	82
China (Takla Makan)*	11	3	2	9	16	16	9	9	5	4	5	2	1	81

Note:
* drift potentials estimated: leaders (. . .), no data.
Source: Fryberger (1979) table 13.

(the maximum value being 489 for the northern An Nafud of Arabia), the intermediate-energy wind environments lying between 200 and 400 units, and the low-energy environments being less than 200. The lowest values are for the Takla Makan of China and the Thar of Rajasthan (82, and 81 vector units respectively).

It is likely, however, that wind energy levels have varied during the course of the Quaternary. Studies of ocean core sediments, for example, prompted Parkin and Shackleton (1973) to suggest that higher trade-wind velocities occurred during glacials.

Conclusions

Although for many decades of this century the power of wind erosion has often been down-played, the presence of extensive yardang fields (often revealed by remote sensing), of large numbers of deflation basins (pans), and of dust storm events has caused a reassessment of ideas to occur. Nonetheless, we are still desperately short of hard quantitative data on rates of rock abrasion or basin excavation by wind. Many of the data we have are for relatively unresistant lake sediments and the like, but there are some limited data for artefactual materials and buildings that suggest that in dry, windy and low-lying situations materials of the strength of brick can abrade very rapidly.

Dust deposition is a phenomenon that can be quantified by direct observation and through the analysis of ocean and ice cores and loess profiles. Large amounts of dust are entrained into the atmosphere, particularly from arid areas, and this represents the operation of a powerful geomorphological process – deflation. Rates of dust deposition were especially high in glacial times, leading to the formation of the world's great loess deposits at rates of between 500 mm 1000 a^{-1} and 3000 mm 1000 a^{-1}.

The movement of dune sand is another palpably important manifestation of aeolian activity. The fastest moving dunes are relatively small isolated barchans. Large barchans and linear dunes move much more slowly. The great ergs are probably ancient features that have taken a long time to evolve, but the speed at which sand moves through a sand sea or erg varies greatly according to prevailing wind energy levels. Some areas have relatively high wind energy conditions compared with others, so that dune morphology, rates of sand transport and rates of aeolian abrasion may vary to a marked degree between different deserts. However, there is some evidence that during glacial times wind velocities in the trades may have been higher than at the present time.

SIX

Glaciers and ice caps

Glacial protectionism and the custard bowl problem

During the nineteenth century, and well into the twentieth, geomorphologists were much exercised by the question of how much erosive work could be achieved by glaciers. Points for discussion included the issue of whether or not glaciers reposed in pre-glacial erosional depressions and tectonic fracture zones, much as, in Ruskin's phrase, custard reposes in a custard bowl, or whether glaciers were hugely effective erosional agents that transformed whole landscapes in a vigorous manner.

One person who had severe doubts about the power of glacial processes to achieve wholesale landscape modification was the durable climber-geologist Bonney. In 1896 (p. 16), for instance, he remarked that 'No one accustomed to travel in non-glaciated as well as in glaciated regions can fail to decipher the familiar characteristics of ordinary rain and river action, though these are sometimes blurred by the palimpsest writing of the ice-scribe.' He believed that the amount of material transported by glaciers as ground moraine was probably exaggerated, and that much of it in any case had been swept beneath a glacier by lateral torrents. He later criticized Davis's cycle of glacial erosion (Bonney, 1902) and suggested that had Davis done 'more than travel through the Alps by one or more frequented routes', and had Davis examined 'the upper as well as the lower parts of valleys' and contemplated 'the mountains from their peaks as well as their bases' he would have had a better appreciation of the limited role of glacial action. It was Bonney's contention that permanent snow is a protective rather than destructive agent. He was also loath to see cirques, such an important part of Davis's block diagrams of the glacial cycle, as being manifestations of severe glacial erosion, and regarded most Alpine valleys as being largely pre-glacial in origin. 'Cirques, corries and bowl-like heads of valleys,' he said, 'are mainly the work of water, their forms depending on local circumstances . . . They are not restricted to glaciated regions . . . cirques occur on the grandest

Plate 6.1
Features like Milford Sound, a great fjord on the west coast of the South Island of New Zealand, appear to give good evidence for the power of glacial erosion. However, considerable controversy still exists as to just how effective glaciers are in comparison with other modes of denudation.

scale where the ice would have the smallest extension, the shortest duration and the least erosive action.'

Another opponent of the idea that glaciers were highly effective erosional agents was Gregory (1913), who made a detailed study of the nature and origin of fjords (plate 6.1). He rejected the idea that fjords were of either glacial or fluvial origin, and championed the view that tectonic processes had been instrumental in their formation. He argued that the fact that most fjords are located in areas that were once glaciated is purely a coincidence resulting from the fact that polar areas have been affected by greater tectonic oscillations than the equatorial zone, and that true fjords exist in low latitudes (e.g. in north-west Australia) where the tectonic history has been suitable. Equally, he pointed out that many glaciated areas do not have fjords. He was particularly exercised by the fact that the shape and pattern of fjords did not square with an origin involving glacial excavation (p. 13):

Study of the fjords system of any country shows that the course of the fjords is inconsistent with the lines of flow of the chief glaciers. The glaciers discharged from the highlands or from the great domes of snow which sometimes formed on the lee side of the existing watershed; and the ice flowed by the most direct channels to the nearest lowlands or to the sea. Many of the fjords had directions which were quite useless to the outflowing ice; and they appear to have been simply filled with stagnant ice, while the main flow of the glaciers was above and across them.

He argued that fjords were pre-glacial features (as Bonney had argued for Alpine valleys) and that they resulted from earth movements associated with the Alpine Orogeny. He suggested they were excavated in the Pliocene (p. 15), 'so that the later ice of the Pleistocene period used the fjords and did not originate them'.

One of the leading exponents of glacial protectionism was Garwood (1910): he asked certain basic questions to which he felt answers should be given (pp. 310–11):

Does an ice-cap lying on a plateau carve the surface more rapidly than the normal weathering agents? Do snow and ice lying on ledges, or resting in gullies, degrade more rapidly than do frost or water? and does a glacier overdeepen its bed more rapidly than would a river flowing in the same valley? No one with any knowledge of glaciated regions doubts that moving ice erodes; that is not now the question. The whole problem turns on the *relative* rate of erosion accomplished by ice as compared with that which would take place over the same district by ordinary erosive agents, namely, weathering, wind, frost, rain, and running water, more especially in a district which is partially covered and partially free from ice.

It may eventually be proved beyond doubt that ice is the greater erosive agent, but this has not been done. It may also be shown conclusively that all the features to be presently described can be satisfactorily accounted for by ice erosion; but so far this is not the case.

Garwood, both a climber and a scientist, as was Bonney, sought to examine some of these problems in the light of his Alpine experiences, and came up with the view that the relative power of ice and river very much depended on where one was – on higher valleys and slopes the role of glacier ice was essentially protective, whereas in the lower portions of valleys where many large affluents coalesced, glacial excavation was more vigorous.

The campaign against glacial protectionism was waged on two main fronts: the morphological evidence provided by impressive erosional and depositional forms, and studies of the amount of material that was being excavated from glaciated catchments in streams draining from them. Shaler (1898), for instance, discussed the turbid nature of glacial meltwaters (p. 222):

A little observation will show the student that this very muddy character of waters emerging from beneath the glacier is essentially peculiar to such streams as we have described. Ascending any of the principal valleys of Switzerland, he may note that in some of the streams flow waters which carry little sediment even in times when they are much swollen, while others at all seasons have the whitish colour. A little further exploration, or the use of a good map, will show him that the pellucid streams receive no contributions of glacial water, while those which look as if they were charged with milk come, in part at least, from the ice arches. From some studies which the writer has made in Swiss valleys, it appears that the amount of erosion accomplished on equal areas of similar rock by the descent of the waters in the form of a glacier or in that of ordinary torrents differs greatly. Moving in the form of ice, or in the state of ice-confined streams, the mass of water applies very many times as much of its energy of position to grinding and bearing away the rocks as is accomplished where the water descends in its fluid state.

However, it was probably the morphological evidence that proved most instructive, as is evident, for example, from this small extract

Plate 6.2
The large quantities of sediment (both in suspended, bed and dissolved form) emerging from glacial snouts (this example is in the Sanetsch area of the Valais in Switzerland) have been used by many workers as evidence for the power of glacial erosion and transport.

from Reclus (1881, p. 120) on the power of a glacier to mould a mountain:

... slowly, but with an invincible force, it works as do the wind, snow, rain, running water, to renew the planet's surface: wherever glaciers have passed over, during one of the ages of the earth's existence, the aspect of the country has been transformed by their action. As do avalanches, they carry the rubbish of the crumbling mountains into the plains, not by violence, but by the patient labour of every moment.

In more recent times a range of methods has been used to measure amounts of glacial erosion. These have included:

1. The use of artificial marks on rock surfaces later scraped by advancing ice.
2. The installation of plattens to measure abrasional loss.
3. Measurements of the suspended, solutional and bed-load content of glacial meltwater streams and of the area of the respective glacial basins (plate 6.2).
4. The use of sediment cores from lake basins of known age which are fed by glacial meltwater.
5. Reconstructions of pre-glacial or interglacial land surfaces.
6. Estimates of the volume of glacial drift in a given region and its comparison with the area of the source region of that drift.

The first four methods apply to present-day glacierized regions and the last two to regions of Pleistocene glaciation. It is by no means obvious that the two different time-scales will lead to comparable results, and that it is possible to use present-day rates of glacial denudation to infer what happened in the past. As Harbor and Warburton (1992, p. 751) have remarked, 'During full-glacial times, ... basins would have higher percentage glacier cover, but also far greater ice thickness, discharges, velocities, basal shear stresses and basal water pressures.' Modern glaciers, the majority of which are decaying and retreating, are not necessarily satisfactory analogues for the vastly more active and substantial glaciers and ice caps of full glacial times.

There is also the problem of knowing how much of the debris transported by glaciers was originally glacially entrained. Much material falls onto glaciers because of non-glacial processes. In some high relief, high altitude areas rates of denudation might be high in the absence of glaciation.

Following his collection of data on stream sediment loads, Corbel (1964, p. 397) was adamant about the importance of glacial erosion:

For several years we have increased the measurements on rivers leaving glaciers and representing glacial erosion. They all have, without exception, some very special characteristics: a fantastic load of mud and pebbles, an almost total absence of dissolved material (the melted ice gives water which is chemically very pure and as pure as distilled water). This chief difference with periglacial torrents proves that it is not simply the result of sub-glacial fluvial erosion, but the result of the melting of ice. The speed of erosion is very superior to that of all non-glacial streams.

He suggested that the rates of erosion for advancing glaciers amount to being paroxysmal – between 10,000 and 50,000 mm 1000 a^{-1}. For glaciers that are stable or in decline the rate may still be 2000 mm 1000 a^{-1}.

In 1968, Embleton and King tried to bring together most of the available information on rates of glacial erosion, including the work of Corbel (1959, 1962). They suggested, though with some caution given the paucity and sparseness of data (p. 251):

The mean rate for erosion by active glaciers probably lies in the range 1000 to 5000 m^3, which should be compared with what we know of rates of fluvial erosion. For the Mississippi, the figure is 50 m^3, for the Colorado above Grand Canyon, 230 m^3, for the Hwang Ho 1000 m^3. Again these figures must not be thought accurate, but they do suggest that glacial erosion is several times more potent than fluvial erosion. [Note: these values are in terms of m^3 km^{-2} a^{-1}]

However, they drew an important distinction between the role of glaciers and the role of ice sheets on low angle surfaces (p. 260):

All methods of estimating glacial erosion are therefore characterized by doubts and uncertainties. But the evidence accumulated so far certainly suggests that active glaciers may be extremely potent agents of erosion, working at many times the speed of rivers of equivalent dimensions in terms of water discharge, while ice-sheets moving sluggishly on surfaces with little regional slope may be comparatively powerless even given the whole span of the Pleistocene.

When in 1975 Andrews came to review the available data, he came to rather different conclusions from Embleton and King.

Indeed, he believed that the view that the mean rate of erosion by active glaciers was in the range 1000 to 5000 mm 1000 a^{-1} was probably wrong by a factor of between 2 and 10! He suggested that a more reasonable estimate was a rate of 50 to 1000 mm 1000 a^{-1}, broadly the same as for normal fluvial catchments. He noted, moreover, that most of the data on rates determined from current meltwater streams are biased in favour of high mountain environments and are largely restricted to valley glaciers that are wet based. Such values, he argued (p. 122), 'cannot be considered typical of values for the erosion of bedrock floors by a large ice sheet'. Certainly, there is little to be gained by comparing glacial erosion in high relief environments with fluvial erosion in more subdued terrain.

Eleven years later, in another textbook, Drewry (1986) again stressed the inadequacy of data on this important subject, and expressed a note of caution about sampling strategies (p. 87):

Significant problems are encountered in estimating total sediment flux (suspended, bed and solution loads). Unless an adequate frequency of regular sampling is undertaken, short-duration, high-magnitude flood events which evacuate considerable quantities of subglacial sediments may go unobserved. Estimates will, therefore, be reduced by a factor of two or three at best and an order of magnitude at worst. In addition not all erosional products are necessarily removed by water – some are entrained in the ice and may never communicate with the glacier hydraulic system.

He presented some published data on erosion rates for selected glaciers based on the measurement of suspended sediment transport (table 6.1a), and suspended and bed-load transport (table 6.1b). The values range from about 73 mm 1000 a^{-1} to 30,000 mm 1000 a^{-1} (with a mean of about 4000 mm 1000 a^{-1}).

Another attempt to summarize available information in the context of some original data based on lake sedimentation rates in southern New Zealand was made by Hicks et al. (1990). They compared, first of all, the sediment yield of Ivory Basin in the Southern Alps, a glacierized basin, with yields from a number of unglacierized basins that are cut in the same lithologies (fissile schist) and lie in the same general climato-tectonic region. They found no appreciable difference. They then incorporated these results into an international dataset of sediment yields for glacierized and unglacierized basins, and found that precipitation causes greater variation in sediment yield than does degree of glacial cover. Summerfield and Kirkbride (1992) found this information convincing and remarked (p. 306), 'The limited data that are available in fact indicate that denudation rates in glaciated catchments are broadly comparable to those in non-glaciated basins with similar local relief and precipitation.'

The use of suspended sediment load data for estimating rates of glacial erosion suffers from all of the problems that are encountered in estimating rates of denudation in fluvial catchments (see

Table 6.1 Erosion rates for selected glaciers

Glacier	Mean basin-wide erosion rate (mm 1000 a⁻¹)
(A)	
Muir, Alaska	19000
Muir, Alaska	5000
Hidden, Alaska	30000
Engabreen, Norway	5500
Storbreen, Norway	100
Heilstugubreen, Norway	1400
Hoffellsjökull, Iceland	2800–5600 [max]
Kongsvegen, Svalbard	1000
St. Sorlin, France	2200
Imat, CIS	900
Ajutor–3, CIS	700
Fedchenko, CIS	2900
RGO, CIS	2500
Bas Glacier d'Arolla, Switzerland	3600
(B)	
Nigardsbreen	165
Engabreen	218
Erdalsbreen	610
Austre-Memurubre	313
Vesledalsbreen	73

Source: various in Drewry (1986) tables 6.2 and 6.3; and Warburton and Beecroft (1993).

chapter 3). As Warburton and Beecroft (1993) have pointed out, with a few notable exceptions, sediment sampling in proglacial steams has been infrequent, sampling frameworks have varied drastically and lengths of observation have been inconsistent. Moreover they assert that the role of subglacial and proglacial storage in modifying the relationship between erosion and sediment yield has rarely been evaluated. They also raise the question as to whether sediment yield data should be standardized for *catchment* area or *glacier* area. Moreover, they point out that while it has been general practice to use meltwater suspended sediment yields to estimate glacial erosion rates, it should also be acknowledged that meltwater bed load, dissolved load and sediment discharge through the ice (i.e. moraine deposition) should also be included (figure 6.1).

Gardner and Jones (1993) estimated rates of denudation for the Raikot Glacier in the Punjab Himalaya. They achieved this by mapping and measuring the variable distribution and thickness of surface debris and sediment concentrations within the ice. They also calculated discharge values. This gave them denudation rates of 4600–6900 mm 1000 a⁻¹ for the glaciated area and 1400–2100 mm 1000 a⁻¹ for the whole basin. These values are of the same order of magnitude as those presented in table 6.1.

In their study of the Bas Glacier d'Arolla in the Valais Region

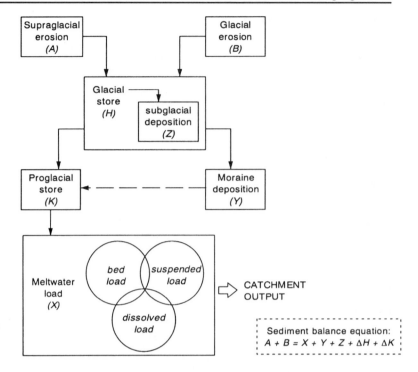

Figure 6.1
*A schema of the glacio-
fluvial sediment transfer
system and sediment
balance equation (source:
Warburton and Beecroft,
1993, fig. 12).*

of Switzerland, Warburton and Beecroft (1993) found that their
calculated rates of erosion differed by almost an order of magnitude depending on the measurements and assumptions made. They
conclude with some recommendations of the ways in which the
calculations of glacial erosion rates can be improved (pp. 26–7):

If meltwater material load data are used to calculate glacial erosion rates
it is recommended that: (1) loads need to be corrected for the effect of
proglacial storage if measured downstream of the snout; (2) meltwater
stream yield estimates should be combined with estimates of moraine
deposition in order to calculate total glacier yield; and (3) erosion rates
should be weighted in terms of source of glacial sediment production.
Though these recommendations should drastically improve estimates, their
accuracy is still limited owing to non-quantifiable parameters – changes
in sub-, en- and supraglacial storage and problems in determining contributions from tributary glacial systems. It should not be overlooked that
some part of the glacier yield is merely transported from supraglacial
zones which undergo frost shatter and avalanches, and this is not strictly
speaking the product of glacial erosion. For progress to be made sediment
budget studies of moraines, proglacial areas and glaciers need to be integrated into a basin-wide glacio-fluvial sediment budget.

Sediment volumes were used by Larsen and Mangerud (1981)
to estimate rates of cirque basin excavation in western Norway
during the Younger Dryas. The length of the erosion period was
determined by radiocarbon dating of associated glaciolacustrine
sediments, and the volume of eroded bedrock was calculated from

measurements of the deposited sediments (including the end moraine, the glaciofluvial delta and the glaciolacustrine sediments).
This produced an erosion rate of 500 to 600 mm 1000 a^{-1}. Were
that rate to be constant during the whole period of cirque formation (which, of course, is most unlikely) it would require at least
83,000 to 125,000 years to erode the cirque to its depth of 50 to
75 m. Reheis (1975) used a combination of suspended load data
and sediment volume data to estimate how long it would take for
the excavation of the Arapaho Glacier cirque in the Colorado
Front Range of the USA, and came up with values that ranged
from between 40,000 and 300,000 years.

An example of the use of stratigraphic methods to determine
long-term rates of glacial erosion is provided by the study of Bell
and Laine (1985). They sought to estimate the degree of erosion
achieved by the glaciation of the Laurentide region of North
America through a study of sediment volumes determined through
seismic studies and Deep Sea Drilling Project boreholes. They
calculated that over a period of about three million years (the
approximate duration of mid-latitude glaciation in the Late
Cenozoic) somewhere between 120 and 200 m of erosion would
be required over the glaciated Canadian Shield to produce the
observed volume of sediment. This is equivalent to a rate of between 40 and 67 mm 1000 a^{-1}. The region was not, of course,
glaciated throughout the whole of the three million year period,
but even so these rates are less than most of the short-term rates
determined from meltwater load studies. Plainly, however, great
care needs to be exercised in using offshore sediment volumes
because of the problems of determining the sediment source and
because of potential problems caused by sea-level changes.

There is, however, considerable debate about just how severely
the Laurentide Shield was eroded by glaciation. On the one hand
workers like White (1972) argue that the ice sheets exposed large
areas of Precambrian basement and carved great basins (e.g. Hudson
Bay), a view that largely coincides with that of Bell and Laine
outlined above. On the other hand, another group of workers
have argued that the depth of glacial erosion achieved on the
shield by late Cenozoic glaciation has only been of the order of
metres or tens of metres. They argue, for example, that much of
the shield would have been an area of low ice velocities, and that
the ice would have been rendered relatively powerless because it
could have been frozen to its bed over much of the shield. These
arguments are concisely reviewed by Braun (1989).

Clearly, geographical location is an important control of the
rate of glacial erosion. Some areas will have had characteristics
that would limit the power of glacial erosion (e.g. resistant
lithologies, low relief, frozen beds), but other areas would have
suffered severe erosion (e.g. non-resistant lithologies, proximity to
former fast ice streams, thawed beds, etc.).

Braun (1989, p. 245) argues that the Appalachians were an area

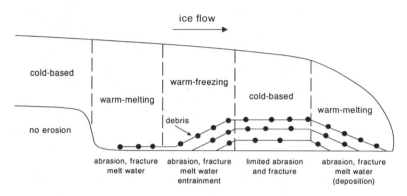

Figure 6.2
Sugden's model of the relationship between styles and rate of glacial erosion and the basal thermal regime of the ice (source: Sugden, 1978).

of active glacial erosion, as revealed by the volume of glaciogenic ocean sediment derived from the area. Using the ^{18}O record from the ocean cores to deduce the duration of glaciation he postulates that rates of glacial denudation could have been as much as 1210–3040 mm 1000 a^{-1}, rates which are more than an order of magnitude greater than estimates of current fluvial erosion in the Appalachians (c.20 to 30 mm 1000 a^{-1}).

Another area of active glacial erosion in the Pleistocene was the western coastline of Norway, an area that is serrated by deep fjords. Using a range of geological and geomorphological techniques, Nesje et al. (1992) have estimated that the overall rate of erosion in the Sognefjord area has been about 610 m over a period of 600,000 years, some of which has been achieved by non-glacial processes. This amounts to an average rate of about 1020 mm 1000 a^{-1}. However, along the fjord itself, the amount of erosion over the same period is considerably greater (i.e around 2000 m), suggesting a glacial erosion rate of about 3300 mm 1000 a^{-1}. This rate is remarkably similar to that determined for the Appalachians.

Sugden (1978) postulated that the erosional style of an ice sheet was very closely related to the basal thermal regime of the ice. Moving outwards from the centre of divergence of an ice cap he identified five idealized zones. Erosion is minimal under the cold-based ice at the centre of the cap because shear from ice flow is limited and the ice is anchored to the underlying bedrock. Surrounding this core is a zone of basal melting where basal slip between ice and rock promotes fracture, plucking and abrasion (figure 6.2). Such a broad pattern is modified at a more local scale by topography, which influences the thickness of the ice layer and also whether it is converging or diverging. Generally an upland which is only thinly covered by ice, and which has caused ice to diverge around it, will tend to be protected beneath cold-based ice. In contrast, depressions tend to channel and carry more ice flow, which favours higher basal temperatures and erosion. Chorley et al. (1984, p. 515) illustrate this with respect to the glacial relief of Scotland, where in the Western Highlands there is a classic

Table 6.2 Rates of glacial abrasion

| | Average abrasion rate | | | |
Locality	Marble plate	Basalt plate	Ice thickness	Ice velocity
Breidamerkurköjull 1 (Iceland)	3 mm a^{-1}	1 mm a^{-1}	40 m	2.6 m a^{-1}
Breidamerkurjökull 2	3.4 mm a^{-1}	0.9 mm a^{-1}	15 m	19.5 m a^{-1}
Breidamerkurjökull 3	3.75 mm a^{-1}		32 m	15.4 m a^{-1}
Glacier d'Argentière	up to 36 mm a^{-1}		100 m	250 m a^{-1}

The error in measurement of average lowering of the surface of the plates to give abrasion rates is approximately 0.3 mm.
Source: from Boulton (1974) table 1.

ice-scoured landscape of fjords and diffluent valleys and troughs, while much more selective erosion occurred in the Cairngorms further east, where pre-glacial tors highlight the degree of protection that has occurred on the plateau tops. Indeed, in the lowlands still further eastwards in Buchan evidence of glacial erosion is sparse, and pre-glacial materials (including deep weathering profiles) have remained largely intact despite having been submerged beneath an ice sheet. In that area glacial erosion achieved no more than 50 m of excavation.

Kleman (1994), in a more general review of the preservation of landforms under ice sheets and ice caps, demonstrated from Fennoscandia and Laurentia that a range of quite delicate land-forms, including eskers, drainage channels and boulder fields had escaped destruction despite complete ice overriding with a duration of several tens of millennia. He believed that this was primarily explicable in terms of the presence of dry (cold)-based ice.

Glacial abrasion

The presence of moulded rock surfaces, grooves and striations, together with large volumes of rock floor in meltwater streams, provides circumstantial evidence for the importance of abrasion at the base of glaciers as a cause of glacial denudation. However, few direct measurements of abrasion have been made. Moreover, current rates may, as already stated, apply to relatively torpid glaciers, while the ability to access only a limited and often marginal section of the glacier bed may reduce the value of such determinations as there are.

The most impressive observations are those that were undertaken by Boulton (1974) in the European Alps near Chamonix and in Iceland. He attached rock and metal plates to bedrock beneath the Glacier d'Argentière and Breidamerkurjokull. The rock types were marble and basalt. The results (table 6.2) varied between 0.9 and 36 mm a^{-1}. They illustrate, to use Boulton's words (p. 48):

That abrasion rates under normal subglacial conditions can be relatively high, that they depend on rock type, ice thickness, and the concentration of debris in the glacier sole. The measured abrasion rates . . . indicate that erosion by this process alone could excavate deep valleys and basins. It is worthy of note that the lowest figure . . . is twice the world average for the erosion of lowland river basins, and that these measured rates would be adequate to produce by abrasion alone, most deep alpine troughs given the lengths of time for which they have been occupied by ice during the Quaternary.

Glacial deposition

In 1974, Goldthwait (p. 164) posed a simple question and then partially answered it:

Glacial deposits and channels are notoriously rapid in their period of actual emplacement or occupation. All eskers, kames, and hillside channels witnessed were completed within one year or a decade. Of course the deposition of stony ablation moraine at any one place was almost diurnal but spreads over new terrain for as long as the ice retreats.

Repetition is commonplace and because of the cyclical expansion and contraction of nearly all glaciers the accumulated deposits become thick. Even the large lateral-loop moraines in and near mountains, tend to be stacked layers of repetitive tills sandwiched with marginal stream gravels, as at Lituya Bay. The topography is such as to impede slightly larger advances, thus compounding and regenerating the stack. Several short periods of deposition may actually span many thousands of years; most of the time record is in hiatuses and in paleosols or buried vegetation.

Two exceptions to the rapid speed of deposition are basal tills and outwashes. Till is widespread, but the till sheet of any one advance and retreat is usually thin and grew at one to five centimetres per year. Except for early deposition in depressions, or spot depression later to be picked up and redeposited, basal till deposition occurs in the last few centuries of glaciation. Outwashes are restricted to gentle gradients. Due to the tendency of depositing streams to fill the channels with bars, to divide, and to flow over recently unoccupied parts of the valley plain, deposition is constantly shifting. Valley trains start long before the ice melts away and continue long after the ice uncovered the area, as long as the annual flood of meltwater feeds the river.

Rates of glacial movement

Glaciers move, advance, retreat, build up and waste away. The purpose of this section is to investigate the rates of these different components of glacial change.

The velocities at which glaciers move are generally in the range of 3 to 300 m a^{-1} for most of their length, but in steep ice falls the velocity can reach 1000 to 2000 m a^{-1} (Chorley et al. 1984, p. 440). Movement at the lower velocities is probably largely by internal deformation processes (e.g. in sluggish, cold-based glaciers), whereas higher velocities reflect the component added by basal sliding. However, a few glaciers flow at rates which are a whole order of magnitude higher. An example of this is the great Jakobshavn Isbrae outlet glacier in Greenland, which flows at 7000 to 12,000 m a^{-1}. Such large outlet glaciers have an immense ice supply to transport and this is sufficient to maintain a fast mode of sliding. There are also some glaciers, surging glaciers, which are prone to periodic surges in their rate of movement in which a wave of ice moves down them at velocities of 4000 to 7000 m a^{-1} (10 to 100 times faster than their previous velocity).

A crucial control of the rate of glacier flow is the relationship between ablation and accumulation (Embleton and King, 1968, p. 34):

In a glacier system that has a high rate of accumulation and also a high rate of ablation, the rate of flow of the ice must be considerable. This applies for example to the Franz Josef glacier, which flows to within about 200 m of sea-level in a relatively low latitude. Its accumulation area lies in an area of very heavy precipitation on the western side of the Southern Alps of New Zealand, while its snout penetrates down into the coastal rain forests. At the other extreme are some of the Antarctic glaciers that receive a very small quantity of precipitation, owing to the extreme cold, and whose only method of ablation is by calving, as temperatures rarely rise to freezing point. These glaciers move extremely slowly. The Ferrar glacier near the Ross Sea, for example, moves at only about 5 cm/day although it is a very large glacier.

Thus, glaciers with a high input of snowfall and a relatively warm climate are much more active than those that have low snowfall inputs and low temperatures. There is a general trend of high to low activity from coastal temperate zones (e.g. western New Zealand) to those of continental and polar zones.

Plainly, the annual net balance between accumulation and ablation varies over a glacier in a systematic manner, with a positive balance (net accumulation) in the upper parts of the glacial system and a negative balance (net ablation) in the lower. These two zones meet (see figure 6.3) at the equilibrium line, where accumulation is exactly compensated by ablation, producing a mass balance which is zero.

Rates of accumulation of snow and ice on glacier surfaces are highly variable in both space and time, and the variability is largely, though not entirely, related to climatic conditions. Thus in Antarctica gross rates of accumulation are low because of the continent's aridity, with values ranging from 2–40 cm water equivalent per annum. By contrast rates are much higher in Greenland, averaging 31 cm water equivalent per year, and exceeding 150 cm in the south-east corner (Embleton and King, 1968; Ohmura and Reeh, 1991).

Figure 6.3
Variations in accumulation and ablation down an idealized glacier. The distribution of net annual accumulation and ablation over a glacier is of great significance to glacier movement. Net ablation generally increases down-glacier below the equilibrium line while net accumulation tends to increase up-glacier above this point. The rate of increase in the net balance with height up a glacier is important for the rate of ice movement because the higher the rate of increase the faster is the rate of ice movement needed to maintain the same glacier profile (source: Summerfield, 1991a, fig. 11.4).

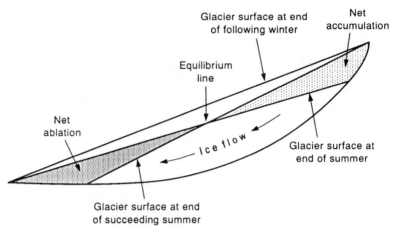

The retreat rate of valley glaciers and ice sheets

Since the nineteenth century, many of the world's alpine glaciers have retreated up their valleys as a consequence of the climatic changes, especially warming, that have occurred in the last hundred or so years since the ending of 'The Little Ice Age' (Oerlemans, 1994). Studies in the changes of snout positions obtained from cartographic, photogrammetric and other data, therefore permit estimates to be made of the rate at which retreat can occur. The rate has not been constant, nor the process uninterrupted. Indeed, some glaciers have shown a tendency to advance for some of the period. However, if one takes those glaciers that have shown a tendency for a fairly general retreat (table 6.3) it becomes evident that as with most geomorphological phenomena there is a wide range of values, the variability of which is probably related to such variables as topography, slope, size, altitude, accumulation rate, and ablation rate. It is also evident, however, that rates of retreat can often be very high, being of the order of 20 to 70 m a^{-1} over extended periods of some decades in the case of the more active examples. It is therefore not unusual to find that over the last hundred or so years alpine glaciers in many areas have managed to retreat by some kilometres.

The rather precise dates for ice-sheet retreat across Scandinavia gained from clay-varve chronology, enable one to calculate the rate at which the great Scandinavian ice sheet wasted away at the conclusion of the Last Glacial. It took about 4000 years to retreat a distance from south to north Sweden of about 1000–1100 km (*c.*0.25 km a^{-1}). The ice mass was over 2 km in thickness, indicating that thinning took place at a rate of about 0.5 m a^{-1}.

The British Ice sheet of the Last Glacial was in excess of 1.5 km thick. It appears to have reached its greatest extent at around

Table 6.3 Retreat of glaciers in metres per year in the twentieth century

Location	Period	Rate
Breidamerkurjökull, Iceland	1903–48	30–40
	1945–65	53–62
	1965–80	48–70
Lemon Creek, Alaska	1902–19	4.4
	1919–29	7.5
	1929–48	32.9
	1948–58	37.5
Humo Glacier, Argentina	1914–82	60.4
Franz Josef, New Zealand	1909–65	40.2
Nigardsbreen, Norway	1900–70	26.1
Austersdalbreen, Norway	"	21
Abrekkbreen	"	17.7
Brikdalbreen	"	11.4
Tunsbergdalsbreen	"	11.4
Argentière, Mont Blanc	1900–70	12.1
Bossons, Mont Blanc	1900–70	6.4
Oztal Group	1910–80	3.6–12.9
Grosser Aletsch	1900–80	52.5
Carstenz, New Guinea	1936–74	26.2

Source: Tables, maps and text in Grove (1988).

Region	Period	Mean rate
Rocky Mts	1890–1974	15.2
Spitzbergen	1906–1990	51.7
Iceland	1850–1965	12.2
Norway	1850–1990	28.7
Alps	1850–1988	15.6
Central Asia	1874–1980	9.9
Irian Jaya	1936–1990	25.9
Kenya	1893–1987	4.8
New Zealand	1894–1990	25.9

Source: Oerlemans (1994).

18,000 years BP (or slightly after) and to have disappeared by 13,000 years BP, a rate of thinning of about 0.3 m a^{-1}.

The British ice sheet retreated northwards from south Wales to Scotland over a distance of c.750 km. Given that this took about 5000 years, the rate of retreat is about 0.15 km a^{-1}.

The great Laurentide Ice sheet did not begin large-scale wasting until c.14,000 years BP, and had more or less shrunk to its present rump by 7000 years BP. Given that it was over 3 km thick, this gives a rate of about 0.42 m a^{-1}.

The rate of retreat of the Laurentide sheet was about 2000 km in 6000–7000 years, a rate of about 0.3 km a^{-1}.

Such rapid rates of ice sheet wastage are made evident in the deep-sea isotopic record (figure 6.4a), and rapid sea-level rise (figure 6.4b) into parts of the Fennoscandian and Laurentide regions could have accelerated the rate of decay brought about by ablation

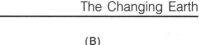

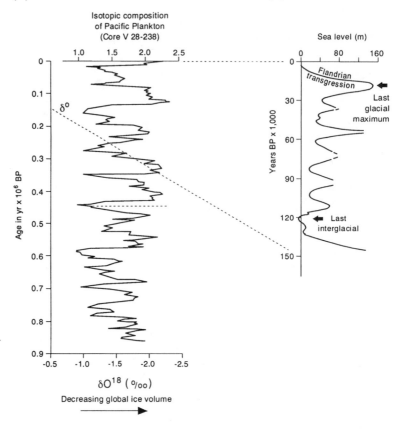

Figure 6.4
The pattern of ice volume change (A) determined by the isotopic examination of a deep-sea core for the last 900,000 years. Note the rapid fluctuations that have taken place in ice volume and the tendency for ice volumes to decay very rapidly at the end of each glacial. This is brought out clearly in the global eustatic sea-level curve for the last 150,000 years (B), which shows the very rapid rise in sea level during the Flandrian transgression as a result of rapid ice cap decay following the Last Glacial Maximum (source: modified from Goudie, 1992, figs 2.5 and 6.9).

because of the role of cliffing and iceberg formation in causing the ice margin to retreat.

Ice sheet accumulation

Given the ready availability of dates for the retreat stages of the great ice sheets, one can have some confidence in the data that have just been presented. The same cannot be said for estimates of the time required for the same ice caps to accumulate. The dates of ice sheet initiation are still relatively imprecise and controversy surrounds the question of how long it takes for major ice caps to develop.

Until the early 1970s the normally accepted view was that the process was a slow one. The limited precipitation totals for high latitude interiors was seen to limit the speed at which build up could occur, and ice cap growth was generally conceived to start from highland massifs and to gradually trundle outwards.

Moreover, at that time it was only just becoming apparent from studies of deep-sea cores that glacial and interglacial fluctuations had been both rapid and frequent. The position was expressed thus by Andrews and Mahaffy (1976, p. 167):

Until the last few years it was tacitly assumed that the development of the large continental ice sheets of the northern hemisphere, specifically the Laurentide and Fennoscandian ice sheets, took many thousands of years. ... estimates of $\geq 15,000$ years were quite acceptable.

For example, Weertman (1964) had used some models to estimate the rate of growth and shrinkage of nonequilibrium ice sheets (p. 157):

The time required to build up an ice-age ice sheet is of the order of 15,000 to 30,000 yr if the accumulation rate over the ice sheet is in the range of 0.2 to 0.6 m/yr. Although there is no way known of obtaining palaeo-accumulation rates, it can be argued that these rates are reasonable in light of what is known about accumulation on the Greenland Ice Sheet.
    ... the time required for the disappearance of an ice-age ice-sheet is smaller than the build-up time provided that the average ablation rate is at least twice the average accumulation rate and that ablation occurs over an appreciable area during shrinkage.

He suggested that ablation rates of 1 to 2 m a^{-1} would lead to shrinkage times of the order of 2000 to 4000 years.

    However, in the 1970s ideas began to change. As already noted, the deep-sea cores suggested rapid changes in ice volumes and sea levels at the end of interglacials. In addition, field work in northern Canada suggested that ice sheet growth might not necessarily follow the mode of gradual outward development from an initial highland core. Furthermore, Lamb and Woodroffe (1970) suggested ways in which the general circulation might have been so modified as to enable rapid snow accumulation to take place, and for what was called 'instantaneous glacierization' to occur. Indeed, Lamb and Woodroffe (1970, p. 33) radically reduced the time they felt was necessary to lead to ice sheet accumulation:

The evidence ... plainly suggests that at least some time between 60,000 and 70,000 years ago ice sheets were quickly established over northern and northwestern Europe, probably including some parts of the British Isles, and over some extensive hinterland in North America (presumably enveloping the Hudson's Bay region), all within 1000–5000 years.

    Andrews and Mahaffy (1976) found this rate excessive, partly because they disagreed on the dates for ice sheet initiation, but even they used various models to suggest that large ice sheets could develop in around 10,000 years under optimum climatic conditions.

Conclusion

For the best part of a century after the development of the glacial theory and the ice age concept in the 1820s and 1830s geomorphologists and other earth scientists debated the power of glacial erosion. The glacial protectionists stressed the importance of long phases of pre-Pleistocene non-glacial erosion to explain alpine valleys, while fjords were sometimes explained in terms of faults, fractures and other structural and tectonic controls.

Even since the availability of sediment load data for glaciated catchments became more substantial in the 1960s there has been debate about how much glacial erosion is being achieved at the present time with views ranging from the paroxysmal ones of Corbel to the more cautious ones of Andrews. Major sampling problems exist and in any event it can be argued that most of the present glaciers on the face of the earth are relatively inactive in comparison with the state they were in during the glacials of the Pleistocene or the Neoglacial advances of the Holocene.

Morphological evidence for estimating rates of long-term glacial modification of landscapes is, however, no less contentious, and there has, for example, been a long-running debate about the extent to which the Laurentide shield area of North America was eroded by glaciation.

Plainly, however, the rate of glacial erosion depends very greatly on a suite of environmental controls so that any naive and simplistic comparison of 'glacial' versus 'fluvial' is doomed to be misleading. Some areas with resistant lithologies, low relief and frozen beds will have low rates of glacial erosion, whereas other areas with non-resistant lithologies, proximity to fast ice streams, rapid snow and ice accumulation, and thawed beds, etc., would be areas with high rates of glacial erosion. Glaciers are efficient transport agents in that they can carry large volumes of big debris provided by non-glacial processes (e.g. rock falls, avalanches), but they also cause some abrasion of their beds and measurements of abrasion rates, though still limited in number, indicate that deep alpine troughs could be created by glacial abrasion in Quaternary times.

It is scarcely surprising that glaciers provided some of the first unequivocal evidence of Quaternary climatic change for they respond very readily to changes in rates of ablation and accumulation. Thus alpine glaciers can retreat at rates of 20 to 70 m a^{-1} in response to the quite modest changes of climate that have taken place since the termination of the Little Ice Ace. At the end of the Last Glacial, the great ice caps thinned at rates of between 0.3 and 0.5 m a^{-1} but retreated at rates of between 0.15 and 0.4 km a^{-1}. Rather less easy to determine is the rate at which great ice caps build up. Traditionally ice caps have been thought to decay rather more rapidly than they build up and there is corroborating evidence in deep-sea cores for rapid deglaciation at the end of the Last Glacial. Values of 15,000 to 30,000 years have often been seen as necessary for an ice cap like the Laurentide or Fenno-scandian examples to develop but markedly shorter time-spans are by no means inconceivable. What is certain is that glaciers and ice caps have expanded and contracted with remarkable frequency during the multiple glacial and interglacial cycles of the Pleistocene.

SEVEN

The coast

Introduction

The sea is one of the most active of geomorphological agents and the coastal zone is thus one of the most dynamic environments. In the nineteenth century there were lengthy debates about the relative power of marine and subaerial processes in moulding the face of the earth, but people have always been interested in the speed at which coastal change occurs, for most of the world's population lives in close proximity to the sea. The dynamism of the sea was graphically stated by Reclus (1873, p. 139):

Although there is necessarily an equilibrium between the work of demolition and that of reconstruction, we would nevertheless, at first sight, be tempted to believe that the sea took the greatest pleasure in destruction. On contemplating the cliffs, those perpendicular walls which on various coasts rise many hundreds of yards above the level of the sea, we are struck with awe to see how the repeated assaults of the waves have been sufficient thus to cut the mountains and hills whose bases were formerly gently sloped to the water. From the top of these cliffs, we see the tumultuous ocean spread at their feet like a plane surface, and we no longer distinguish the billows but by their reflections, or the breakers but by their garland of foam; the multiplied sound of the waves melts into one long murmur, which dies away and rises to die away again. And yet this water, which we see below at such a great depth, and which seems powerless against the solid rock, has thrown down piece by piece all that part of the hill or mountain, of which the cliff is but a gigantic memorial: then, after having thrown down these enormous masses, it has reduced them to sand, and perhaps caused the very trace of them to disappear. Often not even a rock remains where promontories once jutted out. The phenomena ascertained even during the short life of man are facts so grand in their progress, and so remarkable in their efforts, that an English savant, Captain Saxby, has proposed to make of them a special science, *Ondavorology*.

Cliff erosion rates

The denudational power of the sea is best approached through a consideration of the rate at which cliffs respond to Reclus' 'tumultuous ocean'. Cliff recession, being an essentially episodic and localized process associated with storm events and rock falls, is seldom if ever monitored to see the rate of change over a very short time period. Such rates would be very high, whereas more normally workers have measured the long-term average erosion rate for periods of say 10 to 10^2 years. So for example, if a cliff were to retreat by 10 metres in just one storm event that lasted a day, this would, if extrapolated, give an annual rate of 3650 m a^{-1}. However, if such an event only occurs once in every 100 years, the annual rate drops to just 0.1 m a^{-1}. Clearly the frequency and length of observations is crucial.

There are various techniques that can be used to measure rates of cliff recession over longish time spans. The most usual one is to compare the position of the cliff crest using different editions of maps or plans. Also of considerable use are old ground photos, and sequences of air photography. Direct monitoring using stakes and the like can also be attempted.

An excellent review of patterns and controls of rates of cliff recession on a global basis has been provided by Sunamura (1992). He provides a table of rates from many parts of the world, which is far too massive to reproduce here. Employing those data, he finds a general lithological control of the rates of cliff recession over periods that generally last from 1 to 100 years: 10^{-3} m a^{-1} for granitic rocks; 10^{-3} to 10^{-2} m a^{-1} for limestone; 10^{-2} m a^{-1} for flysch and shale; 10^{-1} to 10^0 m a^{-1} for chalk and Tertiary sedimentary rocks; 10^0 to 10^1 m a^{-1} for Quaternary deposits; and 10^1 m a^{-1} for unconsolidated volcanic ejecta.

There are, however, many other controls of the rate of cliff retreat besides lithology. Among these may be degree of exposure, wave climate, frequency of storm surges, the width of the beach fronting the cliff, and the nature of the offshore topography. The influences of the last of these has been well illustrated by Robinson (1980a), who examined the long-term cliff recession history of the Suffolk coast, eastern England. This is a coast composed largely of unconsolidated and unlithified glacial sands and gravels. There is abundant cartographic and archival material which enables records to be extended back to the sixteenth century. The rates were as follows:

1589–1753	1.60 m a^{-1}
1753–1824	0.85 m a^{-1}
1824–1884	1.50 m a^{-1}
1884–1925	1.15 m a^{-1}
1925–1977	0.15 m a^{-1}

This recent marked drop in the erosion rate can be attributed to the amount of wave energy reaching the coast being reduced because of significant divergence in wave refraction, which in turn was produced by a changing bottom configuration associated with the steady northward growth of Sizewell Bank.

The importance of wave climate can be illustrated from further

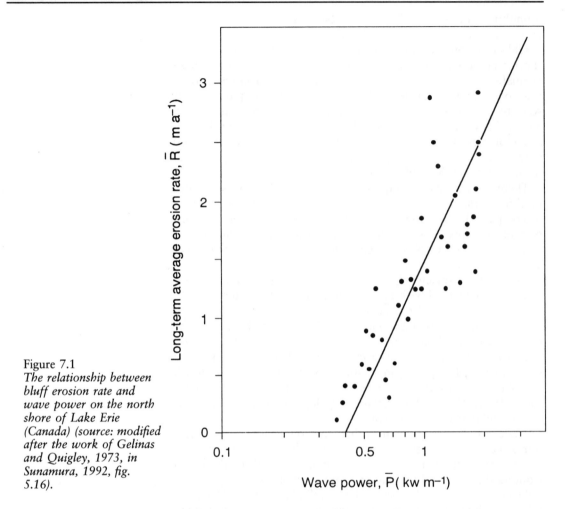

Figure 7.1
The relationship between bluff erosion rate and wave power on the north shore of Lake Erie (Canada) (source: modified after the work of Gelinas and Quigley, 1973, in Sunamura, 1992, fig. 5.16).

north along the eastern coast of England, at Holderness. This is a coastline composed of glacial drift, which extends over a distance of some 55 km between Flamborough Head and Spurn Head. There is a general trend of southward increasing rates of erosion (i.e. towards Spurn Head), and this trend is consistent with increasing wave exposure towards the south. In the north, Flamborough Head offers some protection from severe northerly storms while towards the south an increase in wave height is probably related to an increase in nearshore bottom slope (Robinson, 1980b).

Likewise, Gelinas and Quigley (1973), using long-term cliff recession observations from the shoreline of Lake Erie, where there are glacial materials with no marked alongshore variations in their geotechnical properties, but where there are different degrees of wave exposure, found the clear relationship illustrated in figure 7.1.

There has been some debate as to whether cliff height is a major control of rates of cliff recession. It has been suggested, for example, that large cliffs will produce large amounts of debris which will accumulate at the cliff base, and protect it from further erosion until removed by comminution and longshore transport. This may well hold for short time spans, but the relationship becomes much less pronounced as the time span lengthens (Sunamura, 1992, p. 105).

The general relationship between rates of cliff retreat and lithology can be demonstrated for the coastline of England and Wales (table 7.1). As Goudie (1990, p. 247) expressed it:

As one might anticipate, the highest rates occur on certain rather specific stretches of coastline: the Yorkshire coast between Bridlington and Spurn Head; the Suffolk coast between Sheringham and Happisburgh; the Suffolk coast between Lowestoft and Southwold; the Essex coast at the Naze and Foulness; North Kent between Sheerness and Herne Bay; parts of the chalk coasts of Kent and Sussex; along the Sussex coastal plain to Selsey Bill; the south-west coast of the Isle of Wight; the Hampshire coast between Hurst Castle and Bournemouth; parts of the Dorset coast; Dawlish Warren; the Dee Estuary; and the vicinity of St Bees Head on the Lancashire coast (Bird and May, 1976). It is therefore, with only a few exceptions, the lowland coasts of the east and south where rates of retreat are greatest.

The prime control is lithological, cliff retreat occurring rapidly either where there are unconsolidated glacial drifts or where there are relatively unresistant clays, shales, sands, or chalks. In such materials coast regression may occur at rates in excess of 100 m per century in exposed locations.

In the rocky coastal environments of western Britain the persistence of raised beaches and the presence of composite profiles, many of which show the clear impact of only modestly trimmed Pleistocene cold climate subaerial forms, testify to the slow rates of cliff erosion on hard rock coasts, even in a high-energy wave environment.

Coast recession in North America

Rates of coastal erosion for the USA as a whole, and including cliffed and non-cliffed shorelines, have been plotted by Dolan and Kimball (1985). Their map, reproduced in *The National Atlas of the United States of America*, demonstrates very clearly that retreat is dominant around much of the coastline. This is particularly true of the eastern seaboard, where most of it is retreating at between 0.1 and 2.9 m a^{-1}. There are, however, a few local stretches (e.g. in Georgia and North Carolina) where accretion has been taking place. Conversely there are a few small stretches where the rate of retreat lies between 3.0 and 5.0 m a^{-1}. The Gulf Coast also shows a dominance of retreat, most notably in the vicinity of the Mississippi Delta, and to the west of it. On the west coast of

Table 7.1 Examples of rapid coast retreat

Area	Geology of cliff	Average rate retreat (m 100 a^{-1})
North Yorkshire	Shale	9
North Yorkshire	Glacial drift	28
Holderness	Glacial drift	120
Norfolk		
Weybourne-Cromer	Glacial drift	42
Cromer-Mundesley	Glacial drift	96
Mundesley-Happisburgh	Glacial drift	88
Sratby-Caister	Glacial drift	83
Gorleston-Corton	Glacial drift	57
Pakefield-Kessingland	Glacial drift	105
The Naze (Essex)	Glacial drift, London, Clay and Crag	11–88
Kent		
Reculver	London Clay	68
N Isle of Sheppey	London Clay	96
Isle of Thanet	Chalk	7–22
St. Margaret's Bay-Folkestone	Chalk	7–19
Folkestone	Gault Clay	28
East sussex		
Peacehaven	Chalk	46
Seaford Head	Chalk	126
Birling Gap	Chalk	122
Beachy Head	Chalk	106
Ecclesbourne Glen	Hastings Beds (Sandstone)	119
Fairlight Glen	Hastings Beds (Clays)	143
Cliff End	Hastings Beds (Sandstone)	108
Hampshire		
Christchurch Bay (Highcliffe Castle)	Bracklesham Beds	3
Christchurch Bay (Barton)	Barton Beds	58
Christchurch Bay (Hordle)	Headon Beds	18
Dorset		
Ballard Down	Chalk	23
Kimmeridge Bay	Kimmeridge Clay	39
White Nothe-Hambury Tout	Chalk	21
Ringstead	Kimmeridge Clay	41
Furzy Cliff-Short Lake	Oxford Clay	37
Isle of Wight		
Cranmore	Hamstead Beds	61
Newtown River-Gurnard	Bembridge Beds	38
Brighstone Bay	Wealden Beds	52

Source: Data provided from the works of May, Bird, Robinson and Williams, summarized in Goudie (1990) table 8.1.

the USA rates of change appear to be markedly less, which probably reflects the fact that it is in a very different tectonic and
lithological situation to the eastern and southern coasts. Thus rates
of retreat are nearly always less than 1 m a^{-1}. Some accretion has
been taking place in southern California (between Santa Barbara
and San Diego) and also along the southern coast of Washington
state, but even in these areas there are locations where some cliff
retreat is taking place (see, for example, Orme, 1991) at rates of
up to 0.5 m a^{-1}.

Taking the USA as a whole, and dividing it up on the basis of
different shoreline types (table 7.2), Dolan and Kimball found that
the pattern of rates was for the most part predictable. Coasts with
fine-grained sediments, deltas and mudflats that offer low levels of
resistance to wave attack have the highest mean erosion rates
(c.2.0 m a^{-1}). Most sandy beaches and barrier islands erode at
slightly lower rates (c.0.8 m a^{-1}). By contrast the rock shorelines
of the Atlantic coast show rates of accretion of as much as 1.0 m
a^{-1}.

Because of the presence of a sea ice cover for the cold months
of the year, high latitude beaches may be protected from wave
erosion for substantial periods of time. In such low-energy environments one might intuitively anticipate that rates of coastal
recession would be modest. However, this is not necessarily the
case. Indeed, the coastal bluffs of the Arctic Lowlands of North
America appear to be undergoing rapid retreat that locally exceeds
5 m a^{-1}, and which for significant portions of the coast exceeds 2
m a^{-1} (table 7.3). This rate is higher than the rate of coastal retreat
encountered along the Gulf Coast of the USA (the location with
the highest rate of coastal erosion in the coterminous United States).
Moreover, the erosion of the Arctic lowlands is accomplished
entirely within the three-month open-water period (L. C. Carter et
al., 1987). The explanation for this seems to lie in the presence of
ice-rich unconsolidated bluffs which are subject to a whole suite
of essentially subaerial processes, which include surface wash, debris
slides, ground-ice slumps and thermoerosional falls. Thermoerosion
involves the melting of ground ice and the mechanical removal of
thawed sediment by waves or turbulent currents. Lewkowicz (1987)
provides a detailed analysis of the role of thermokarst slumping
on Banks Island, northern Canada, where the headwall retreat
rates on slumps averages 8.6–11.4 m a^{-1}.

Shore platform erosion

The recession of sea cliffs will tend to lead to the formation of
gently sloping shore platforms at their feet. However, such platforms have a polygenetic origin, in that it is not just wave attack
that is responsible for their development. Thus the old term 'wave

Table 7.2 Rates of coastal change in the USA

A Rates of change for states and regions

Region	Mean m a^{-1}*	Standard deviation	Total range*	N**
Atlantic Coast	−0.8	3.2	25.5/24.6	510
Maine	−0.4	0.6	1.9/−0.5	16
New Hampshire	−0.5	–	−0.5/−0.5	4
Massachusetts	−0.9	1.9	4.5/−4.5	48
Rhode Island	−0.5	0.1	−0.3/−0.7	17
New York	0.1	3.2	18.8/−2.2	42
New Jersey	−1.0	5.4	25.5/−15.0	39
Delaware	0.1	2.4	5.0/−2.3	7
Maryland	−1.5	3.0	1.3/−8.8	9
Virginia	−4.2	5.5	0.9/−24.6	34
North Carolina	−0.6	2.1	9.4/−6.0	101
South Carolina	−2.0	3.8	5.9/−17.7	57
Georgia	0.7	2.8	5.0/−4.0	31
Florida	−0.1	1.2	5.0/−2.9	105
Gulf of Mexico	−1.8	2.7	8.8/−15.3	358
Florida	−0.4	1.6	8.8/−4.5	118
Alabama	−1.1	0.6	0.8/−3.1	16
Mississippi	−0.6	2.0	0.6/−6.4	12
Louisiana	−4.2	3.3	3.4/−15.3	106
Texas	−1.2	1.4	0.8/−5.0	106
Pacific Coast	−0.0	1.5	10.0/−5.0	305
California	−0.1	1.3	10.0/−4.2	164
Oregon	−0.1	1.4	5.0/−5.0	86
Washington	−0.5	2.2	5.0/−3.9	46
Alaska	−2.4	2.0	2.9/−6.0	69
Delaware Bay				
New Jersey	−1.9	1.3	0.3/−3.0	13
Delaware	−1.3	2.1	5.0/−3.0	12
Chesapeake Bay	−0.7	0.7	1.5/−4.2	136
Western shore	−0.7	0.5	1.5/−1.9	67
Maryland	−0.7	0.3	−0.1/−1.3	35
Virginia	−0.8	0.7	1.5/−1.9	32
Eastern Shore	−0.7	0.8	0.1/−4.2	69
Maryland	−0.8	0.9	−0.3/−4.2	47
Virginia	−0.5	0.4	0.1/−1.2	22

* Negative values indicate erosion; the positive values indicate accretion.
** Total number of 3-min grid cells over which the statistics are calculated.

cut platform' is probably misleading, and is based on a neglect of other formative processes, including a wide range of weathering mechanisms. Thus it may be that the primary process leading to their formation is wave-induced cliff retreat, and such cliff retreat lowers the cliff-platform junction surface level. Secondary lowering of the platform is achieved by three categories of action: mechanical, chemical and biological. Mechanical processes include

Table 7.2 (Continued)

B Rates of change for coastal landform types

Region	Mean m a⁻¹*	Standard deviation	Total range*	N**
Mud Flats				
Fla.	−0.3	0.9	1.5/−1.5	9
La. Texas	−2.1	2.2	3.4/−8.1	84
All Gulf	−1.9	2.2	3.4/−8.1	93
Rock Shorelines				
Atlantic	1.0	1.2	1.9/−4.5	36
Pacific	−0.5	–	−0.5/−0.5	7
Pocket Beaches				
Atlantic	−0.5	–	−0.5/−0.5	9
Pacific	−0.2	1.1	5.0/−1.1	144
Sand Beaches				
Marine–Mass.	−0.7	0.5	−0.5/−2.5	17
Mass-NJ	−1.3	1.3	2.0/−4.5	22
Atlantic	−1.0	1.0	2.0/−4.5	39
Gulf	−0.4	1.6	8.8/−4.5	121
Pacific	−0.3	1.0	0.7/−4.2	19
Sand beaches with rock headland	−0.3	1.9	10.0/−5.0	134
Deltas	−2.5	3.5	8.8/−15.3	155
Barrier Islands				
La.-Texas	−0.8	1.2	0.8/−3.5	76
Fla-La.	−0.5	1.7	8.8/−4.5	82
Gulf	−0.6	1.5	8.8/−4.3	158
Maine–NY	−0.3	2.6	4.5/−1.5	12
NY–NC	−1.5	4.5	25.5/−24.6	159
NC–Fla.	−0.4	2.6	9.4/−17.7	256
Atlantic	−0.8	3.4	25.5/−24.6	421

* Negative values indicate erosion; the positive values indicate accretion.
** Total number of 3-min grid cells over which the statistics are calculated.
Source: Dolan and Kimball (1985) tables 1 and 2, with modifications.

the abrasive action of wave-moved sediment, potholing and rock disintegration caused by wetting and drying, frost and salt attack. Chemical processes include the solution of platform-forming rocks, especially carbonates, while biological processes include the action of rock-boring and rock-browsing organisms. It is probable that in most examples all these different processes combine to different degrees to produce any observed platform morphology and rate of development.

In recent years our appreciation of the rates at which platforms develop has been greatly expanded by the use of the micro-erosion meter (MEM), though other methods are available, including detailed repeat photography and other forms of direct instrumentation.

Table 7.3 Coastal retreat rates for the Arctic Lowlands of North America

Location	Mean rate (m a⁻¹)
Chukchi sea coast	
Near Barrow	1.9–2.2
Barrow to Peard Bay	0.3
Alaskan Beaufort Sea Coast	
Near Barrow	2.1
Oliktok Point	1.4
Near Prudhoe Bay	1.0–2.0
Barrow to Harrison Bay	6.3
Smith Bay to Prudhoe Bay	2.5
Barrow to Demarcation Point	3.0
Canadian Beaufort Sea Coast	
Kay Point	1.3
Yukon Territory	.0
Tent Island, Yukon Territory	15.0
Garry Island, NWT	1.2–2.3
Pelly Island, NWT	6.3
Hooper Island, NWT	1.5
Oullen Island, NWT	9.2
Tuktoyaktuk Peninsula	4.0

Source: modified after Carter et al. (1987) table 5, from various sources.

One limitation of the MEM technique is that it cannot measure the rate of change where plucking or quarrying of large rock fragments or jointed blocks are involved.

There have also been attempts to model the rate of platform lowering in relation to the rate of cliff retreat. For example, Zenkovich (1967, p. 168) related the lowering rate dz/dt to the rate of horizontal cliff retreat dx/dt by

$$\frac{dz}{dt} = \frac{dx}{dt} \tan \beta$$

where $\tan \beta$ is the gradient of the bedrock profile of the platform. Although this relationship includes the assumption that the overall nearshore profile shows parallel retreat it gives a reasonable approximation to average lowering rates over a long period (Sunamura, 1992, pp. 125–6). Various empirical field studies indicate that the ratios of vertical lowering of the cliff base to horizontal cliff recession are of the order of 2 to 5 per cent.

There are now many field data on rates of platform development, especially in the intertidal zone. Most rates generally average between 500 and 2000 mm 1000 a⁻¹ (Spencer, 1988, p. 260), but given that a disproportionate number of studies come from limestone coastlines, these figures may exaggerate the real average rate for all lithologies. Certainly, there is a general tendency for the rate of lowering to increase as the hardness of bedrock diminishes (Sunamura, 1992, p. 126). Indeed in unconsolidated rocks,

Table 7.4 Measured vertical erosion rates on shore platforms in temperate areas using the micro-erosion meter

Location	Substrate	Environment	Author	Rate (mm 1000 a^{-1})
N. Adriatic	Limestone	Intertidal	Torunski (1979)	630
Co. Clare (Ireland)	Limestone	Intertidal	Trudgill et al. (1981)	200
S. Island (N. Zealand)	Limestone	Intertidal	Kirk (1977)	380–1350
N. Yorks (UK)	Shales	Intertidal	Robinson (1977)	1000
Victoria (Australia)	Greywackes + siltstone	Intertidal	Gill and Lang (1983)	370
S. Devon (UK)	Greenschist	Supratidal	Mottershead (1982)	610

like the Tertiary London Clay from eastern England, rates may range as high as 10,000–1,000,000 mm 1000 a^{-1}.

The rates of platform development for non-tropical areas shown in table 7.4 are for temperate locations. However, along some colder coastlines, such as the Norwegian, there are very well developed platforms, termed *strandflat*, and there have been frequent suggestions that some of the coastal platforms in areas like Scotland may be periglacial relicts rather than interglacial forms. Some support is given to this idea by studies undertaken on the shorelines of a moraine-dammed lake in southern Norway (Dawson et al., 1987). The rock shores of the lake showed platforms that had been produced in a 75–125 year period at an average rate of 2.6 to 4.4 cm a^{-1}. This figure refers to the rate of widening rather than to the rate of lowering, and so cannot be compared with the data in tables 7.4 and 7.5. Nonetheless, it does indicate the power of frost processes in causing platform retreat and shows that under the right conditions the sorts of platforms found in Scotland at various heights above current sea-level, could have been created (i.e. widened to the extent of 100 m or more) in a few thousands of years.

A particularly useful study of rates of platform lowering in a tropical context was undertaken on Grand Cayman Island in the West Indies by Spencer (1985). Studying limestone substrates and employing the MEM technique he found that overall erosion rates averaged 990 mm 1000 a^{-1}. However, he also found that aspect was highly important. The mean erosion rate on open coasts was 2770 mm 1000 a^{-1}, a rate six times higher than on reef protected shores (450 mm 1000 a^{-1}). Table 7.5 provides some data on vertical erosion rates for tropical shore platforms.

Bioerosion is a highly important process on platforms in many environments, not least in the tropics. Spencer (1988) summarizes copious data on the rate at which various boring organisms can attack rock surfaces, including clionid sponges, polychaetes and bivalve molluscs. Also important is the grazing activity of various echinoids, fish, chitons and gastropods. However, as Spencer (1988, p. 271) remarks, 'Attempts to quantify bioerosional activity are still fragmentary and have yet to reach the stage where broad

Table 7.5 Measured vertical erosion rates on shore platforms in tropical and reef environments

Location	Substrate	Environment	Erosion rate (mm 1000 a⁻¹)	Author	Method
(a) Tropical environments/reef substrates					
Aldabra Atoll	Reef limestones	Intertidal	600–4000	Trudgill (1976)	Micro-erosion meter technique
			1500–2700	Viles and Trudgill (1984)	Re-measurements of Trudgill's (1976) sites
Grand Cayman	Reef limestones	Subtidal	1120–1790	Spencer (1985a, 1985b)	Micro-erosion meter technique
			460	Spencer (1983)	Weight-loss tablet method
		Intertidal	880–1230	Spencer (1985a, 1985b)	Micro-erosion meter technique
			560	Spencer (1983)	Weight-loss tablet method
			6400	Warthin (1959)	Soluble/insoluble blocks
Grand Bahama	Reef limestones	Intertidal	2500	Warthin (1959)	Soluble blocks
Bermuda	Reef limestones	Subtidal	1300	Bromley (1978)	Rock pedestals
Puerto Rico	Reef limestones	Intertidal	1000	Kaye (1959)	Dated surfaces
Norfolk Is. and W. Australia	Aeolianite and beachrock	Intertidal	600–1000	Hodgkin (1964)	Erosion pegs/dated surfaces
Heron Is.		Intertidal	500	Stephenson (1961)	
Lizard Is., Great Barrier Reef	Beachrock Coral	Subtidal	630–1260	Kiene (1985)	Slices of coral *Porites*, upper surface lowering
		Intertidal reef flat	6–20	Kiene (1985)	
	Coral	Lagoon	220–930	Kiene (1985)	
Bikini Atoll	Coral Beachrock	Intertidal	300	Revelle and Emery (1957)	

generalizations about the temporal and spatial patterns of substrate modification by macro-organisms can be made. At present, few rates of bioabrasion are directly comparable referring, within sites, to individuals rather than populations of particular species and, across sites, to varying exposures and lithologies.'

It is also important to remember that some marine organisms can encrust rock surfaces and thus protect them from physical erosion by wave activity. Such protective organisms include some mussels, encrusting barnacles, calcareous algae, vermetids, serpulids, mat-forming organisms and sand binders (including filamentous algae) (Trudgill, 1985, p. 153). On some exposed coasts in Grand Cayman, Spencer (1985, p. 68) found that 'bio-erosion has to compete against bioconstructional processes with rocksealing accretions becoming thicker and better lithified with greater wave energy'.

Coastal progradation and accretion: salt marshes

The erosion of a coastline, so graphically described by Reclus in the introduction to this chapter, inevitably produces the where-

Table 7.6 Estimates of net annual accretion rates for specified marsh ages

Age of marsh (years)	Predicted net annual accretion rate (mm 1000 a⁻¹)
10	17000
100	4800
200	1200
300	280
400	70
500	20

Source: Pethick (1981) table 3.

withal for progradation elsewhere. As Reclus himself put it (1873, p. 158):

But if the sea demolishes on one side, it builds up on the other, and the destruction of the ancient shores is compensated for by the creation of new coasts. The clays and limestones torn from the promontories, the shingle of every kind which is alternately thrown up on the shore and swept back in the waves, the heaps of shells, the silicious and calcareous sands formed by the disintegration of all these fragments, are the materials employed by the sea for the construction of its embankment, and the silting up of its gulfs.

Several methods have been used, some of them over many years, to determine rates of marsh accretion in gulfs and elsewhere. In the short term, rates can be obtained by using natural or artificial marker horizons or repeat survey. In the longer term, they can be determined using archaeological evidence or by calculating age/height relationships (e.g. Pethick, 1981) based on archival information or isotopic dating. These can give dates for marsh segments of different ages.

In general one might anticipate that as a marsh becomes older and higher so the rate of accretion will decline because of less frequent sediment inputs from tidal inundation and because of compaction effects. In his study in North Norfolk, Pethick (1981) did indeed find that net annual accretion rates declined as marshes became older and higher (table 7.6). The relationship between sedimentation rate and height of the marsh is confirmed by the more recent study undertaken by French (1993), also in north Norfolk (figure 7.2).

Another probably important control of accretion rates is tidal range. Harrison and Bloom (1977), working in Connecticut, USA, found a significant correlation between the average rate of accretion and the local mean tidal range. The simplest hypothesis to explain this is that a greater volume of water, carrying suspended sediment, flows over those sites where the tide range is greatest. However, other possibilities exist: the productivity of some sediment trapping plants (e.g. *Spartina alterniflora*) shows a direct linear

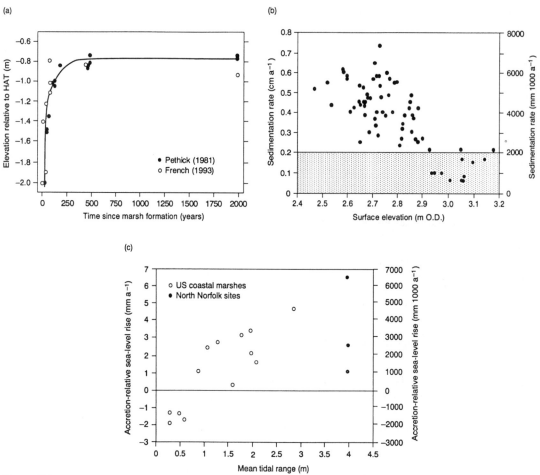

Figure 7.2
Some relationships between accretion rate and environmental factors for coastal salt marshes: (a) The asymptotic increase in mean surface elevation relative to HAT (Highest Astronomical Tide) with increasing marsh age in North Norfolk (employing data of Pethick, 1981 and French, 1993); (b) The annual sedimentation rate versus elevation for Hut Marsh, Scolt Head Island, North Norfolk (mean for 1983–9). The shaded area delimits those points experiencing a net accretionary deficit over this period, assuming regional subsidence and a eustatic rise which together amount to 2.1 mm yr⁻¹ (French, 1993, fig. 2); (c) Relationship between marsh sedimentary balance (net vertical accretion – local sea-level rise) and mean tidal range (French, 1993, fig. 4).

correlation with tidal range, as may faecal pellet production by deposit- and filter-feeding invertebrates.

However, as French (1993, p. 78) points out, the linear relationship (figure 7.2c) between net sedimentation surplus and increasing tidal range may be obscured within highly minerogenic marshes which exist in close proximity and experience similar tidal regime and sediment supply conditions, but which are of widely differing ages.

The nature of a marsh's vegetation community is also an

Plate 7.1
In some areas, especially those with high tidal ranges, high inputs of sediment, fast rates of organic accumulation, and vegetation that is efficient at trapping sediment (in this case from southern England, Halimione portulacoides *along creek margins and* Spartina anglica *on the marsh surface itself), rates of salt marsh accretion can be rapid and are likely to enable marshes to build up should sea levels rise as a result of the 'Greenhouse Effect'.*

important control of the rate of marsh accretion. Large perennial grasses (e.g. *Spartina* spp) may be more effective at trapping sediment than low-growing annuals (plate 7.1).

Table 7.7 summarizes some published results of studies of salt marsh accretion. Values range from zero to as much as 100,000 mm 1000 a⁻¹. The high Bridgwater Bay value is associated with an area of high tidal range, in a heavily turbid estuary, with widespread growth of *Spartina*.

However, it is evident from French's work that rates of sedimentation vary markedly over short distances. As he remarks (1993, p. 65), 'Detailed measurements obtained within the western marshes of Scolt Head Island indicate that hydraulic gradients related to the creek network impart considerable local variability in

Table 7.7 Rates of salt marsh accretion

Location	Source	Rate (mm 1000 a^{-1})
Connecticut, USA	Harrison and Bloom (1977)	2000–6600
North Norfolk, UK	Pethick (1981)	0–17000
Scolt Head, UK	French and Spencer (1993)	1000–8000
Pollen Island, NZ	Chapman and Ronaldson (1958)	300–1700
Severn Estuary, UK	Allen and Rae (1988)	5100
Bridgwater Bay, UK	Ranwell (1964)	80000–100000
Poole Harbour, UK	Ranwell (1964)	20000
Maine, USA	Anderson et al. (1992)	6200–7000
Essex, UK	Reed (1988)	5000–14000
Scheldt, Netherlands	Oenema and De Laure (1988)	4000–15000
Dyfi estuary, Wales	Shi (1993)	4000–15000

Table 7.8 Salt marsh vulnerability

Less Sensitive
Areas of high sediment input
Areas of high tidal range (high sediment transport potential)
Areas with effective organic accumulation

More sensitive
Areas of subsidence
Areas of low sediment input (e.g. cyclically abandoned delta areas)
Mangroves (longer life cycle, therefore slower response)
Constraint by sea walls, etc. (nowhere to go)
Microtidal areas (rise in sea-level represents a larger proportion of total tidal
    range)
Reef settings (lack of allogenic sediment)

sedimentation rate, such that there is as much variation over a distance of 25–50 m as exists over an entire marsh surface'.

The rate at which salt marshes can accrete is of some importance with respect to how they might cope in the face of an accelerating rate of sea-level rise caused by global warming. As table 7.8 indicates, some marshes will be sensitive to change, whereas others will be much less sensitive.

Reed (1990) suggests that salt marshes in riverine settings may receive sufficient inputs of sediment that they are able to accrete sufficiently rapidly to keep pace with projected rises of sea-level. Areas of high tidal range, such as the salt marshes of the Severn Estuary in England/Wales, are also areas of high sediment transport potential and may thus be less vulnerable to sea-level rise. Likewise, some vegetation associations, e.g. *Spartina* swards, may be relatively more effective than others at encouraging accretion, and organic matter accumulation may itself be significant in promoting vertical build up of some marsh surfaces. For marshes that are dependent upon inorganic sediment accretion, increased storm activity and beach erosion, which might be associated with the greenhouse effect, could conceivably mobilize sufficient sediments in coastal areas to increase their sediment supply.

One particular type of marsh that may be affected by anthropogenically accelerated sea-level rise is the mangrove swamp. As with other types of marsh the exact response will depend on local setting, sources and rates of sediment supply, and the rate of sea-level rise itself. However, mangroves may respond rather differently from other marshes in that they are composed of relatively long-lived trees and shrubs, which means that the speed of zonation change will be less. Woodroffe (1990, pp. 484–5) suggests, for example, that on a shoreline where the tidal range is a modest 60 cm, 'and where a monospecific mangrove forest occupies the upper half of that tidal range, a sea-level rise of the order of 30 cm by 2050 would completely displace the mangrove zone within a timespan less than the expected life history of individual mangrove trees. It will take time for mangrove species to regenerate at the levels which are suitable for them at the new stage of sea; in some cases the dead or stressed mature trees from the previous sea-level will shade out or inhibit regrowth of the new saplings'. Woodroffe believes that the degree of disruption is likely to be greatest in microtidal areas, where any rise in sea-level represents a larger proportion of the total tidal range than in macrotidal areas.

The setting of mangrove swamps will be very important in determining how they respond. River dominated systems with a large allochthonous sediment supply will have faster rates of shoreline progradation and deltaic plain accretion and so may be able to keep pace with relatively rapid rates of sea-level rise. By contrast in reef settings, in which sedimentation is primarily autochthonous, mangrove surfaces are less likely to be able to keep up with sea-level rises.

This is the view of Ellison and Stoddart (1990), who used Holocene stratigraphic studies from low islands to establish the historical evidence for the rates of sea-level rise with which mangroves could contend in such sediment-limited environments. It was their view that low island mangrove ecosystems (mangals) had in the past been able to keep up with a sea-level rise of up to 8–9 cm 100 a^{-1} (800–900 mm 1000 a^{-1}) but that at rates of over 12 cm 100 a^{-1} (1200 mm 1000 a^{-1}) they had not been able to persist. On this basis they concluded (p. 161):

The predicted possible rates of greenhouse-induced sea-level rise of 100–200 cm/100 years make it inevitable that most mangals will collapse as viable coastal ecosystems. This implies that mangrove ecosystems of low islands will be more vulnerable to rising sea-level than those of high islands and continental shores.

The growth of reefs

The ultimate exemplification of the importance of bioaccumulation processes in coastal geomorphology is the development of coral reefs. Moreover, given uncertainties of the way in which reefs might respond to future rates of sea-level rise, estimation of their rates of accretion has become an increasingly important issue. In considering the phenomenon of reef growth it is necessary to recognize a distinction between the growth rates of individual organisms, net framework production and reef accretion as a whole (Stoddart, 1990, p. 526). The study of rates of growth can be approached in one of two main ways: retrospectively, by looking at the history of reef growth over centuries and millennia, and actualistically by measurement and calculation of on-going reef growth.

The retrospective methods have been applied to reef growth during the Flandrian transgression of the Holocene. Stoddart shows that most recorded rates have been in the range of 1000–8000 mm 1000 a^{-1}, with maximum values reaching as much as 12,000 mm 1000 a^{-1} under optimal conditions, but falling to much less with decelerating and stabilized sea-levels. Certainly, during the most rapid phases of Holocene sea-level rise (i.e. between c.7000 and 11,000 years ago) reefs were extremely active in terms of accretion, and on the Huon Peninsula, Papua New Guinea, rates of accretion reached 13,000 mm 1000 a^{-1} (Chappell and Polach, 1991).

There may be some regional differences in rates of reef growth as determined by retrospective methods. Adey (1978), for example, suggested that total vertical Holocene reef accretion in the Caribbean had been about twice that reported for the Pacific atolls. Such factors as wave energy differences and different subsidence history may help to explain why the rates vary.

The actualistic methods, which include construction of carbonate budgets and transfers, indicate rates of broadly the same order of magnitude, with values ranging from less than 1000 mm 1000 a^{-1} to as much as 10,000 mm 1000 a^{-1}. Stoddart gives an estimated mean active reef accretion rate of 7000 mm 1000 a^{-1}.

Buddemeier and Smith (1988) took a pessimistic view of the response of coral reefs to rapidly rising sea-level. Employing 15 mm a^{-1} as the probable rise of sea-level over the next century, they suggested that this would be (p. 51) 'five times the present modal rate of vertical accretion on coral reef flats and 50% greater than the maximum vertical accretion rates apparently attained by coral reefs'. Using a variety of techniques they believe (p. 53) 'the best overall estimate of the sustained maximum of reef growth to be 10 mm a^{-1} . . .'. They predict (p. 54) that 'inundated reef flats in areas of heavy seas will be subjected to progressively more destructive wave activity as larger waves move across the deepening flats . . . Reef growth on the seaward portions of the inundated, waveswept reef flats may therefore be negligible compared to sea-level rise over the next century, and such reef flats may become submerged by almost 1.5 m'.

However, reef accretion is not the sole response of reefs to sea-level rise, for reef tops are frequently surmounted by small islands (cays and motus) composed of clastic debris. Such islands might be very susceptible to sea-level rise. On the other hand were warmer seas to produce more storms, then the deposition of large amounts of very coarse debris could in some circumstances lead to their enhanced development.

In vertical section, coral reef growth is strongly controlled by water depth, which controls the amount of light that is available for photosynthesis. Reef corals can grow from the surface down to water depths that receive about 1 per cent of the surface irradiance. In practice this means that the lower limit of coral growth is at about 90–100 m (Bosscher and Schlager, 1992).

Low latitude carbonate depositional environments include depositional mechanisms besides true corals. For example, in sheltered lagoonal environments stromatolites are highly important in causing sedimentation and accretion. They appear capable of rapid upward growth, with figures of as much as 160 mm 1000 a^{-1} being estimated for the Bahamas (Paull et al., 1992). However, rapid growth is far from universal, and by contrast, Playford and Cockbain (1976), working in the classic site of Hamelin Pool in Shark Bay, Western Australia, found that the maximum observed rate of stromatolite growth (over a five-year period) was about 1 mm a^{-1}, but that in most cases there was no observable growth. Indeed, net erosion had occurred in some cases. They also showed that roadways cut through the stromatolites in the 1930s had not been recolonized after four decades.

Delta progradation

Deltas are highly dynamic features of coastlines. Some display very rapid rates of progradation, whereas others show symptoms of decline and retreat. They form as a result of a competition between marine and terrestrial forces, so that if rates of sediment supply from inflowing rivers are higher than the rate at which material is eroded away by the sea, the delta may succeed in accreting, provided that a third controlling factor, subsidence, is not excessive. If rates of wave and current attack are high, if river sediment loads are modest, and if subsidence is active, then retreat may occur. Moreover, some parts of a delta may prograde, while other parts start to erode, and this state of affairs often results from changes in the relative importance of different distributaries (as in the case of the Mississippi Delta).

In favourable circumstances, as for example when a large, sediment-laden river enters a lake or sheltered sea, progradation can occur with sufficient speed to have caused major changes in coastline configuration within historical times. Some data are presented in table 7.9. It is probably not exceptional for rates of progradation to be of some tens of metres per year.

Some of the world's major deltas appear to be suffering from retreat at the present time, and this may in some cases result from anthropogenic disturbance. The damming of rivers (e.g. the Nile

Table 7.9 Rates of delta progradation

Delta	Source	Rate (m a⁻¹)
Fraser R., Canada	Kenyon and Turcotte (1985)	2.3–8.5
Rhine, L. Constance	"	50
Mississippi (SW Pass)	"	76
Hwang-Ho, China	Kukal (1990)	268
Kalantan	"	100
Irrawadi, Burma	"	46–61
Mekong, Vietnam	"	61
Rhone, France	"	30
Yangtze-Kiang, China	"	23
Chao Phraya, Thailand		5–6
Po, Italy		25–129

Figure 7.3
The extending distance and rate of the Yellow River Delta, China, since 1194. The number beside each line segment refers to the extending rate in the corresponding period in m a⁻¹. The concave upward rate reflects the effects of levée construction (source: Jiongxin Xu, 1993, fig. 7).

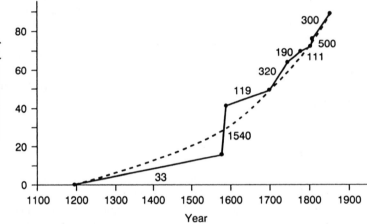

at Aswan and elsewhere, or the Colorado in the USA) may so reduce the amount of sediment transported to the river's mouth that it starts to retreat. Likewise, channelization and levée construction may reduce the amount of sediment that is available to cause accretion in the salt marshes between distributary channels (as in the case of the Mississippi). However, in some cases the cause of retreat may be entirely natural, and result from the current rates of sea-level rise or from changes of cyclic nature that affect different distributaries.

Conversely, human actions have accelerated the rate of growth of certain deltas. A good illustration of this is provided by the Yellow River in China (figure 7.3). The construction of levées, particularly following the efforts of Pan Jixun in the second half of the sixteenth century, has narrowed the width of the river's channel, and increased its flow depth and velocity. As a result the river's sediment carrying capacity has been increased, and there has been a change in the ratio of the amount of sediment deposited on the flood plain to that poured into the sea. More sediment has become available to contribute to delta accretion, leading to the observed increasing delta extending rate with time.

Glacio-eustatic change

Coastal retreat and accretion are not the only important features of a changing coastline. Also important is the rate at which sea level can change.

The waxing and waning of the great ice sheets and glaciers during the glacials and interglacials of the Quaternary, led to substantial changes in sea level on a global (eustatic) basis. At times when water was held in the great polar ice caps and global ice volumes were approximately three times greater than now, sea levels were probably 100 to 120 m lower than today. Estimates vary, however, with opinions generally ranging within 60 m either way of 120 m (Bloom, 1983). Were melting of the present polar ice caps to be complete, global sea levels would rise a further 66 m.

The last glaciation seems to have reached its global maximum at around 18,000 years ago. Shortly thereafter ice caps decayed rapidly, and save for the blip associated with the Younger Dryas, retreated progressively so that by 7000 years ago most of the great Laurentide ice sheet of North America had disappeared. The chronology and rate of ice sheet retreat is discussed in chapter 6. This rapid deglaciation led to the rapid Flandrian transgression of the Holocene. One of the best available records of its progress is provided by the study of radio-carbon dated samples drilled from submerged coral reefs on the Barbados Shelf in the Atlantic (Fairbanks, 1989). The coral that was involved, *Acropora palmata*, grows within a few metres of sea level and thus permits a relatively accurate assessment of past water depths. The oldest and deepest of the samples from Barbados indicates that at the time of the last glacial maximum the maximum depth of the shoreline was at −120 m. Sea level began to rise and accelerated so that during the 1000-year period between 12,500 and 11,500 the rate was as high as 24 m 1000 a^{-1}. The rate decelerated during the Younger Dryas, and then accelerated again until 8000 years ago, when the shoreline was at a depth of around −25 m. Since around 5000 years ago, the rate of sea-level rise has flattened out, which is not surprising given that by that time the great Pleistocene ice caps of Laurentia and Scandinavia had effectively contracted to their present limits. The rates of rise when the transgression was at its most rapid are impressive (20–75 mm a^{-1} round about 8000 years ago), and may have been associated with periods of near catastrophic collapse of ice caps and ice shelves (Tooley, 1993, p. 99). This is the interpretation given by Tooley (1978) to account for a 7 m rise in sea level in just 200 years in Lancashire, north-west England.

Table 7.10   Trends of relative sea-level rise in different parts of the USA

	Average trend (mm a⁻¹) 1940–86
North East Coast	2.68
South East Coast	2.28
Gulf Coast	2.89
South West Coast	1.55
North West Coast	0.12

Source: data of Hicks and Hickman in Fletcher (1992).

During periods of very speedy rise, the sea must have advanced laterally at an extraordinarily rapid rate. In the Perso-Arabian Gulf region, for instance, there was a shoreline displacement of around 500 km in only 4000–5000 years, a rate of no less than 100–120 m a^{-1}. Comparable rates of transgression for the North Sea are 60 km 1000 a^{-1}, 15 km 1000 a^{-1} for the Bristol Channel, and 18–32 km 1000 a^{-1} for the Gulf of Mexico (Evans, 1979).

There is certain evidence to suggest that following on from the waning of the Little Ice Age, and associated with the warming that has characterized much of the last century, sea levels have been rising. Most of the evidence comes from tide-gauge records, and it is very difficult to separate current tectonic submergence and other factors from the eustatic effects of current glacial melting. The rates determined over the last century or so by different workers are summarized in table 6.5 in Goudie (1992), and range from 12–55 cm 100 a^{-1} but it is important to recognize that when Pirazzoli (1989) made allowance for some of these other factors, the rate was diminished to just 4 to 6 cm per 100 a^{-1}.

Because of local factors, the rate of present-day sea-level change will vary from area to area. This is brought out in the case of the United States of America (table 7.10), where there appear to be major differences between different parts of the country (Fletcher, 1992). The rates of relative sea-level rise range from as much as 2.89 mm a^{-1} along the Gulf Coast, to only 0.12 mm a^{-1} for the northern West Coast. These data are for the period from 1940 to 1986.

Figure 7.4 (from Shennan, 1992) shows the most recent attempt to develop a spatial picture of the trends of crustal change in the British Isles. It is apparent from this that the area may be split into two provinces, one in the south-east of England, where there is a tendency for sea-level rise, and the other in the north and north-west where sea-level is falling. These two provinces correspond to the Late Quaternary pattern of glacio-isostatic loading, with large ice masses developed over the north and west and a forebulge over the extreme south and east of England. Rates of subsidence in south-east England approach 2 mm per year in the Thames Estuary and along the southern coastline of East Anglia. By contrast, rates of uplift in western Scotland approach a similar value. However,

(a)

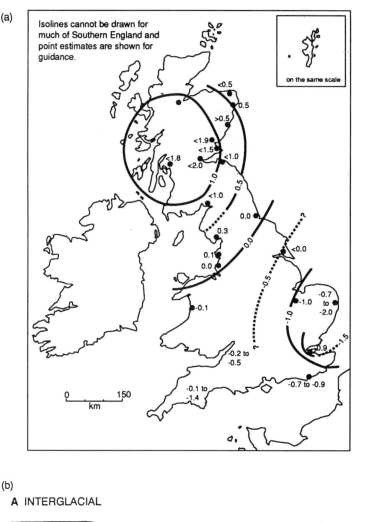

Isolines cannot be drawn for much of Southern England and point estimates are shown for guidance.

on the same scale

<0.5
0.5
>0.5
<1.9
<1.5
<1.8
<2.0
<1.0
<1.0
0.0
0.3
<0.0
0.1
0.0
-0.7
to
-2.0
-1.0
-0.1
-0.2 to -0.5
-0.9
-0.1 to -1.4
-0.7 to -0.9

0   150
km

(b)

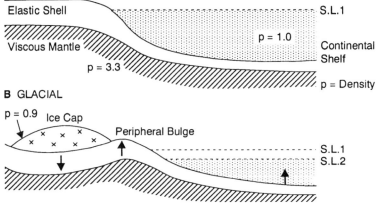

A INTERGLACIAL

Elastic Shell   S.L.1

Viscous Mantle   p = 1.0   Continental
                                                     Shelf
p = 3.3   p = Density

B GLACIAL

p = 0.9 Ice Cap   Peripheral Bulge

                                                     S.L.1
                                                     S.L.2

Figure 7.4
(a) The rate of current crustal movements in Britain.
(b) The glacio-isostatic model of the effects of glaciation on the crust, showing depression under an ice mass and the development of a compensatory fore-bulge away from the ice mass. When the ice mass is removed, the peripheral bulge collapses, causing subsidence to occur, while uplift occurs under the area that was formerly under the ice (sources: Shennan, 1992, fig. 5 and Goudie, 1992, fig. 6.6).

tectonic influences are beyond the scope of this chapter and are something that will be returned to in the next chapter.

If global warming accelerates as a result of the greenhouse effect caused by anthropogenic emissions of carbon dioxide, methane, CFCs, nitrous oxide and other gases into the atmosphere, it is possible that these current rates of sea-level rise will accelerate. This is an area where there has been considerable debate in recent years, associated with considerable uncertainties as to the degree of temperature change that is likely to occur, and the rate at which glaciers and ice caps will change. If, however, we take a consensus figure of about 0.5 m in 100 years during the next decades, then plainly this is very substantially faster than has been experienced in the past century. It is around four times faster than the rates determined by most workers, and ten times faster than Pirazzoli's rate of 4 to 6 cm in 100 years. This whole issue has been discussed by the Intergovernmental Committee on Climate Change (1990) (Houghton et al., 1990), who discount some of the more extreme predictions that sea-level rise could exceed 3.5 m by the year 2100.

EIGHT

Tectonics

Introduction

Thus far in this book the concern has been with phenomena and processes that can be termed *exogenic* (literally, originating from the outside). The rates at which such phenomena operate cover an enormous spectrum of values (see the lower part of figure 8.1), and range, for example, from very slow rates of solution on low relief surfaces developed on resistant substrates, to very rapid mass movements such as rockfalls and debris flows. The other great class of phenomena and processes are those known as *endogenic* (literally, originating from within), and comprise various types of movement or event related to tectonic activity and crustal change. Some generalized data on rates of such endogenic activity, based on the work of Summerfield (1991a), are shown in the upper part of figure 8.1.

The morphology of landscapes depends on the relative speeds of endogenic and exogenic phenomena, and this assertion is borne out by the words of Bloom (1978, p. 19):

Tectonic and volcanic landscapes can be built at rates at least an order of magnitude more rapidly than erosion can destroy them. If intensities were otherwise, our terrestrial landscape would be very dull indeed.

Techniques for assessing tectonic rates

The study of rates of tectonic processes has gained great momentum in the last three decades because of the arrival of the plate tectonic paradigm, and because of a widespread interest in neotectonics – the study of late Cenozoic deformation. A large range of techniques has now been developed to permit an assessment of rates of both vertical and horizontal movements (Doornkamp and Han Mukang 1985). Some of these techniques are based on long-term evidence and some on short-term evidence (table 8.1).

One of the simplest methods for inferring the distribution and

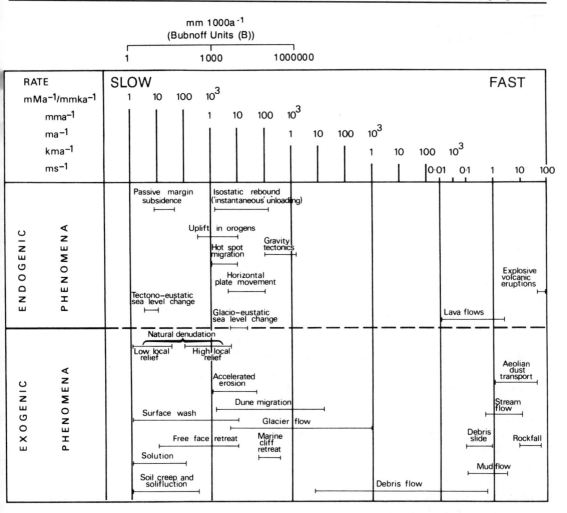

Figure 8.1
Comparison of rates of various endogenic and exogenic geomorphological processes (source: Summerfield, 1991a, fig. 15.1, with minor modifications).

rates of tectonic change is by present-day or historical accounts of specific seismic events. Although there are inevitable problems with eye-witness reliability after a major catastrophe, and although in some cases earthquake-induced landslide scars may be confused with earthquake-induced fault traces, careful analysis of observations can be a productive source of information.

Long-term changes in emergence or submergence of land can be ascertained using archaeological remains of known age (see, for example, Flemming, 1969). The classic example of this approach was provided by the temples at Pozzuoli in Italy, where Babbage (1847) was able to show, from analysis of the local geology and molluscan borings in some of the columns of the Temple of Serapis,

Table 8.1 Techniques for determining rate of tectonic movement

	Nature of movement
Long-term	
Altitude of dated marine sediments or landforms (e.g. coral reefs, beach rock)	Vertical
Altitude of archaeological remains (e.g. harbour installations)	Vertical
Deformation of sediments of known age (e.g. cave sediments, fans, terraces, raised beaches and platforms)	Horizontal
Palaeomagnetism of submarine basalts	Horizontal
Historical analyses of earthquake records	Vertical
Short-term	
Paired gauges on lakes	Tilt
Tide gauge levels	Vertical
Geodetic levelling	Vertical
Direct observations of earthquake traces	Vertical
Electronic distance measurement and laser ranging	Horizontal
Gravity changes	Vertical
Tiltmeters and creepmeters	Horizontal

that the ruins had subsided into the sea and then been gradually uplifted. Examination of the displacement of archaeological sites can also demonstrate faulting, as in the case of the Great Wall of China (Deng et al., 1984) and the Hisham Palace in Jericho (Reches and Hoexter, 1981). Evidence of recent anticlinal uplift is demonstrated by the long profiles of Iraqi canals, while arched river terraces may be dated by archaeological means, e.g. at El Asnam in Algeria (Vita-Finzi, 1986).

Various other dating techniques have also been employed with some success to date tectonic events and degrees of change. Thus faults, which can be identified on the ground or through remote sensing, can be related to geomorphological features of known age.

It is desirable to use precise chronometric dating methods when these are available. In California, Sieh and Jahns (1984) used the charcoal present in two ages of alluvial fan and two ages of channel alluvium to estimate the ages of channel offsets of measured length caused by Holocene movement along the San Andreas Fault. Similarly, Clark et al. (1972) used the degree of deformation of dated lake deposits that were originally broadly horizontal as the basis for reconstructing an episodic fault displacement chronology for the last 3000 years in the Salton Sea area. Likewise tree-ring analysis may permit the identification of trees that have been stressed by fault movement, and may also provide a means whereby such stress can be dated (Sieh, 1978). Lichenometric dating may be applied to fault scarps (Nikonov and Shebalina 1979). On a longer time-scale, Williams (1982) employed an ingenious geomorphological method for estimating uplift in New Zealand's South Island, where Quaternary marine terraces sometimes cut across limestones that contain caves formed in association with past sea-level-controlled groundwater levels. Uplift, followed by water-table lowering and abandonment of cave passages by active streams, permitted speleothem deposition. Dating of these deposits by ^{230}Th/^{234}U and by ^{14}C gives a minimum age for the cave levels and hence also for any terrace to which they may be related, while successive dates at different levels yield an uplift rate for the area. Even longer-term rates of uplift can be obtained by fission-track analysis of minerals like apatite, as has been attempted by Zeitler et al. (1982) in the Karakorams of Pakistan, and by Miller and Lakatos (1983) in the Adirondacks of the USA. Such thermochronologic techniques only yield direct data on rates of rock uplift relative to the surface (i.e. denudation). In themselves they provide no information on vertical movements of either rocks or the surface with respect to the geoid.

When we discuss uplift it is important to recognize that it can be used in two distinct senses (Summerfield, 1991a, p. 371). *Surface uplift* refers to the upward movement of the landsurface with respect to a specific datum, which is normally sea level. *Crustal uplift* refers to the upward movement of the rock column with respect to a specific datum, which is also normally sea level. Crustal uplift will only equal surface uplift if there is no denudation during the period of uplift. If, as one would expect, denudation does occur, then the crustal uplift rate will exceed the surface uplift rate. If the rate of crustal uplift is exceeded by denudation, then there will be no surface uplift and surface elevation will be reduced.

Warping of geomorphological features of known age may also be identified. The classic example of this is the suite of warped shorelines caused by hydro-isostatic forces around Pleistocene Lake Bonneville (Crittenden, 1967). However, the tilting of marine raised beaches of Pleistocene age may also be inferred from their morphology, because they may be steeper than modern platforms and because the discrepancy becomes greater with age (Bradley and Griggs, 1976).

Recent vertical crustal movements (RVCM) are those which have been occurring during the present century and have been determined instrumentally (see, for example, Pavoni and Green, 1975; Vyskocil et al., 1983). The most common techniques for their investigation involve geodetic levelling and relevelling over transects or networks. This can provide data on rates of subsidence or doming over periods of some decades, and can help to explain certain aspects of fluvial morphology (Burnett and Schumm 1983). It is important, however, to recognize some of the limitations and inaccuracies of some past geodetic data (Bilham and Simpson, 1984).

Tide gauge records provide a means to determine rates of vertical movement with respect to sea level, though allowance has to be made for such factors as eustatic changes and changes in water level brought about by meteorological and other circumstances. Tidal analyses have been used extensively (see, for example, Lisitzin 1974; Grant 1980) because gauges are in existence in many parts of the world. Similar techniques can be applied to lakes, the extent of tilting over a period of time being obtained by comparing water levels on pairs of gauges at different parts of a lake basin (Wilson and Wood, 1980).

A whole range of new survey techniques is permitting the determination of phenomena such as geoidal changes and movements of plates. Satellite altimetry now enables continuous monitoring of the geoid (Horam, 1982) while laser ranging is beginning to supply measurements of interplate movement. Laser ranging can also be used with respect to satellites, and Smith et al. (1979) have derived the distance between two points on opposite sides of the San Andreas Fault by laser tracking of near-Earth satellites as part of an experiment to estimate the motion along the plate boundary. Very-long-baseline interferometry (VLBI) can also be employed (Shapiro 1983; Kroger et al., 1986).

Faults and folds of vigorous activity are tempting subjects for direct gauging with such devices as wire-type creepmeters and electronic tiltmeters. Tiltmeters can be installed in boreholes, but may be adversely affected by rainfall events and by groundwater fluctuations (Edge et al., 1981). Such techniques, nonetheless, provide a useful means for determining rates of movement along faults (Bolt and Marion, 1966; Johnston et al., 1976).

A full appreciation of current ground deformation probably requires the use of a variety of techniques in combination. For example, Chadwick et al. (1983) used five different methods to monitor ground deformation associated with the eruption of Mount St Helens:

(a) trilateration and triangulation around the volcano to monitor changes in shape;
(b) long-ranging distance measurements to points in the crater made from a point 8.5 km from the lava dome;
(c) measurements of short horizontal and vertical distances across active cracks and thrust faults on the crater floor;
(d) slope-distance and vertical-angle measurements from sites on the crater floor to targets on the floor and dome; and
(e) measurement of ground tilts on the crater floor by means of inexpensive and expendable electronic tiltmeters and precise levelling.

Sea-floor spreading

The great geological discovery of the early 1960s was that the process of continental drift might be accomplished by a process named sea-floor spreading (Dietz, 1961). It was suggested that at the ocean ridges upwelling and partial melting of material from the asthenosphere occurs, thereby creating new oceanic lithosphere. With the progressive creation of this lithospheric material, the oceans widen and the continents marginal to the ocean are moved apart. Thus, the drift that has separated Europe from the Americas has been accomplished by the gradual opening up of the Atlantic Ocean. However, the increase in size of those oceans that are growing by sea-floor spreading from their ridges is largely balanced by lithospheric destruction in other oceans that are shrinking. Such shrinking is achieved by subduction of material at deep sea trenches situated around their margins. This is, therefore, a fundamental consideration for understanding geomorphology at the global scale.

The process of subduction produces new continental crust through volcanic activity. The volcanoes occur primarily either within a continental setting (as in the Andes) or in an oceanic setting (as with the New Hebrides Arc). Howell (1989, p. 33) estimates that if the growth rate of a volcanic arc, averaged over a normal life span of 20–80 Ma, is of the order of 20 to 40 km^3 Ma^{-1} for each linear km of arc, this equates to a world-wide rate of creation of new sialic material of about 1 km^3 a^{-1}.

The mean age of the oceanic crust is surprisingly modern – c.55 Ma (Howell, 1989, p. 52), indicating a rate of recycling of about 110 Ma (figure 8.2). Very little of the sea floor is older than Jurassic and this indicates vividly the significance of global tectonics as a process that has shaped the gross morphology of the Earth's surface.

By dating the new lithospheric material formed at the central oceanic ridge, and relating its position to that of the ridge from which it originated, it is possible to estimate with some reliability the long-term rates of sea-floor spreading. The prime dating method that has been employed is magnetostratigraphy, underpinned by potassium argon dating of the lavas that hold the magnetic signal. Table 8.2 is a compilation of the spreading rates that have been calculated for mid-ocean ridges from around the world.

These rates of spreading range from a low of 6 mm a^{-1} in the northern part of the North Atlantic, to rates of over 60 mm a^{-1} in parts of the Pacific Ocean. For what it is worth, an average value for all 29 locations is just under 25 mm a^{-1}.

The faster moving plates (Pacific, Nazca, Cocos and Indian) have the common feature that a large fraction of their perimeters is being subducted, whereas, in contrast, the more slowly moving plates (American, African, Eurasian and Antarctic) have large continents embedded in them and do not have significant attachments of downgoing slabs. Press and Siever (1986, p. 504) put forward an hypothesis to explain this which 'Associates rapid plate motions with the "pull" exerted by large-scale downgoing slabs, and slow plate motions with the "drag" associated with embedded continents' (figure 8.3).

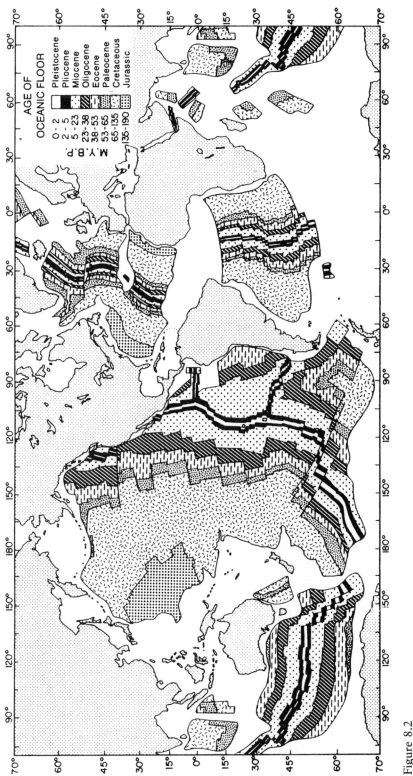

AGE OF
OCEANIC FLOOR

Pleistocene		0 - 2
Pliocene		2 - 5
Miocene		5 - 23
Oligocene		23 - 38
Eocene		38 - 53
Paleocene		53 - 65
Cretaceous		65 - 135
Jurassic		135 - 190

M.Y.B.P.

Figure 8.2
The age of the sea floor, indicating that there is no sea floor older than the Jurassic (source: Allegré, 1988, fig. 29).

Table 8.2 Spreading rates at mid-ocean ridges ('spreading rate' is defined as the accretion rate per ridge flank)

Ridge	Latitude	Observed rate (mm a^{-1})
Juan de Fuca	46.0°N	29
Gulf of California	23.4°N	25
Cocos–Pacific	17.2°N	37
	3.1°N	67
Galapagos	2.3°N	22
	3.3°N	34
Nazca–Pacific	12.6°S	75
Chile Rise	43.4°S	31
Pacific–Antarctic	35.6°S	50
	51.0°S	44
	65.3°S	26
North Atlantic	86.5°N	6
	60.2°N	9.5
	42.7°N	11.5
Central Atlantic	35.0°N	10.5
	23.0°N	12.5
Cayman	18.0°N	7.5
South Atlantic	38.5°S	18
Antarctic–S. America	55.3°S	10
Africa–Antarctic	44.2°S	8
NW Indian Ocean	4.2°N	14
	12.0°S	18.5
	24.5°S	25
SE Indian Ocean	25.8°S	28
	50.0°S	38
	62.4°S	34.5
Gulf of Aden	12.1°N	8
	14.6°N	12
Red Sea	18.0°N	10
Mean		25

Source: data from various sources in Kearey and Vine (1990) table 4.1, with modifications.

However, there has been some debate in the literature as to whether plates containing a significant amount of continental crust move as fast as oceanic plates. Forsyth and Uyeda (1975) suggested that because of excess asthenospheric drag on continental plates they would tend to move more slowly than oceanic plates, and that, furthermore, plate velocity would vary inversely with the amount of continental material. More recently, one should point out, Meert et al. (1993) have suggested that continental plates are in fact capable of moving at rates which are not dissimilar to those achieved by oceanic plates, and they use palaeomagnetic data for Laurentia and Gondwana to indicate that in the past large continental plates have travelled faster than 16 cm a^{-1}.

The great pioneer exponent of continental drift, Alfred Wegener, envisaged that rates were very much greater than those determined by recent studies. In his *Die Entstehung der Kontinente und Ozeane* (1915) he mistakenly used the similarity of glacial moraines on

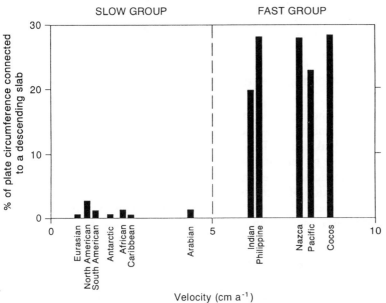

Figure 8.3
The relationship between plate speeds and the percentage of the plate circumference connected to a descending slab (source: Allegré, 1988, fig. 9.1).

both sides of the North Atlantic Ocean to suggest that North America and Europe had still been linked in the ice ages of the Quaternary. This would imply an implausibly rapid rate of drift since the waning of the great northern hemisphere ice sheets. Indeed, he was willing to speculate that Greenland and Europe were now separating at a rate of over 10 m a^{-1}.

Related to these rates of sea-floor spreading are the rates at which linear volcanic chains and aseismic ridges form as oceanic crust moves over relatively stationary magma sources called 'hotspots'. Chains of seamounts, guyots and volcanic isalnds in the Pacific, many of them the foundations of coral atolls, have been interpreted by the hotspot model. For instance, radiometric dates demonstrate that the focus of volcanism in the Hawaiian chain has migrated to the south-east towards Kilauea at a linear rate of about 10 cm a^{-1} for the last 30 million years. Similar linear decreases in the age of volcanism occur towards the south-east in the Marquesas, Society and Austral islands in the South Pacific, with rates of migration of the order of 11 cm a^{-1} and in the Pratt-Welker seamount chain in the Gulf of Alaska at a rate of about 4 cm a^{-1}.

These linear chains are, however, only examples of the interaction between hotspots and the rate of ocean crust movement. If oceanic plate movements relative to hotspots are small, large quantities of magma are intruded or extruded, forming a large island or lava plateau with thick crust (e.g. Iceland). If, on the other hand, plate hotspot motion is erratic in rate and direction, irregular clusters of volcanoes may form (e.g. possibly, the mid-Pacific mountains) (Condie, 1989, p. 158).

Horizontal displacements along faults

Where the earth's crust is broken by active faults, horizontal displacements may occur, either gradually, or suddenly during individual events. Table 8.1 shows some of the methods that are available to determine rates of crustal movement, both horizontal and vertical. For example, Sieh and Williams (1990) used the deformation of dated sediments to estimate the rate of slow horizontal creep along the southern portion of the San Andreas fault in California, and came up with rates of 2–4 mm a^{-1}. The question of the use of palaeomagnetism of basalts is discussed in the section on rates of sea-floor spreading.

Kukal (1990, table 3, p. 16) provides a list of horizontal slip rates that have been determined along prominent fault systems. These are reproduced in modified form in table 8.3. Kukal (1990, p. 15) concluded from these data:

. . . the rate of horizontal movement varies in time and space. It is an episodic movement in which active periods alternate with quiet periods. Various branches of the same fault system may move at different rates, and even various points of the same fault can display different rates of movement.

The rates of displacement vary from tenths of millimetres to tens of millimetres per year. Many faults display rates of more than 10 mm per year, the maximum observed being 110 mm per year.

The distinctly episodic character of the displacements is responsible for the rates measured over shorter intervals of time being higher than the long-term values.

In a sense Kukal is wrong with regard to the latter point. If movements along faults are episodic, as they probably are, short-term observations may fail to detect infrequent or episodic movements, and so short-term observations may underestimate long-term rates. This was certainly the view of Knuepfer (1992), who worked on the rates of right-lateral slip movement along the major faults of the South Island of New Zealand. By dating stream terraces and moraines, and looking at the degree of displacement that had taken place across features of differing ages, he showed that rates over the last three to five thousand years have been very much less than the rates for the Late Quaternary as a whole.

It is also interesting to note that in addition to the New Zealand rates of fault movement being variable and episodic through time, they also showed considerable variability in space, even though the faults were all part of the great Alpine Shear System. Latest Quaternary right-lateral slip rates varied from 3–4 mm a^{-1} for the

Table 8.3 Seismotectonic horizontal slip rates along prominent fault systems

Fault/location	Slip rate [mm a^{-1}]	Time span
San Andreas, California	max. 85 ave 10–20	since the Cretaceous
	12 ± 3.9	Pliocene–Quaternary
Hayward, California	7	Pliocene–Quaternary
Furnace Creek, California	0.25	Pliocene
San Jacinto, California	1–4 total displacement 38 km	
San Andreas, California	35.8 ± 5.4	last 13,250 years
	33.9 ± 5.4	last 3700 years
San Gregorio, California	6–13	Late Pleistocene–Holocene
San Gregorio-Hongri, California	6–13	Pliocene–Quaternary
Calaveras, California	3–10	Pliocene–Quaternary
Concord, California	3–10	Pliocene–Quaternary
Grenville, California	0.1–0.7	Pliocene–Quaternary
Basin Ranges, USA	7	Late Pleistocene–Holocene
Denali, Alaska	10–20	since the Late Cretaceous
McKinley, Alaska	35	Quaternary
Alpine, New Zealand	10	Holocene
	14–18	Recent
Shorya-Oki	10	Recent
Sanriku, Japan	40	Recent
Tokaido, Japan	40	Recent
Surkob, Pamir, USSR	15–20	Recent
Red River, Yunan, China	2–5	Quaternary
Upper Rhine, W. Germany	0.5	Recent
Lower Rhine, W. Germany	2.0	Recent
Rhine, Graben, Belgian zone	0.7	Recent
Swabian Jura	3	Recent
North Anatolia, Turkey	31–110	Quaternary
Greater and Lesser Antilles	2.0 ± 0.5	Historical times
Gulf of Suez, North Egypt	3.0	Present
Al Asnam, Algeria	0.76	Historical times

Source: modified after Kukal (1990) table 3.

Porter's Pass Fault to 25–40 mm a^{-1} for the Hope Fault (Kneupfer, 1992).

Vertical movements of the crust

As a general rule, the vertical movements of the crust, be they in the form of either subsidence or uplift, are generally an order of magnitude smaller than horizontal movements. Vertical movements occur on a global scale, both at plate boundaries and at plate interiors. At plate boundaries, major uplift may be associated with the tectonics of continent–ocean or continent–continent collision, while within plate interiors movements may be caused by a range of mechanisms: by lateral variations in the thermal regime, by the response of the crust to variations in surface loading caused by the erosion of elevated regions and the deposition of sediments away

Table 8.4 Long-term mean uplift rates in orogenic zones determined by various methods

Location	Method	Rate (mm 1000 a⁻¹)	Period
Central Alps	Apatite fission track age	300–600*	6–10 Ma BP
Central Alps	Rb and K-Ar apparent ages of biotite	400–1000*	10–35 Ma BP
Kulu-Mandi Belt, Himalayas	Apparent Rb-Sr ages of coexisting biotites and muscovites	700*	25 Ma BP – present
Southern Alps, New Zealand	Apparent K-Ar ages of schists	10,000*	1 Ma BP – present
Southern Alps, New Zealand	Estimated ages of elevated marine terraces	5000–8000**	140 ka BP – present
Huon Peninsula, Papua New Guinea	U-series and ¹⁴C dating of elevated marine terraces	1000–3000**	120 ka BP – present
Bolivian Andes	Fission track dating	100–200*	20–40 Ma BP
Bolivian Andes	Fission track dating	700*	3 Ma BP – present

* Rock uplift
** Surface uplift
Source: from various sources in Summerfield (1991a) table 15.1, and in Benjamin et al. (1987).

from these regions, or by the vertical response of the crust to horizontal forces (Lambeck, 1988).

Orogenic uplift

Some of the very fastest rates of uplift occur where plate collision takes place, causing crustal shortening. Summerfield (1991a, p. 375) (table 8.4) has summarized the situation as it relates to some of the world's major mountain ranges:

Minimum crustal uplift rates in major orogenic belts, such as the Alps and the Himalayas, located at convergent plate margins range from 300 to about 800 m Ma^{-1} averaged over periods of several million years. But rates may be much higher than this; in southern Tibet, for instance, Late Pliocene–Early Pleistocene terrace deposits containing a fauna indicative of a lowland sub-tropical climate have since been elevated to a height of 4000–5000 m indicating surface uplift averaging more than 2000 m Ma^{-1}. Overall, crustal uplift rates in the Himalayas appear to be currently averaging around 5000 mm yr^{-1}. This is matched by long-term rates in some Andean Ranges, such as the Cordillera Blanca in Peru which, on the basis of the exposure of a granite batholith emplaced at a depth of 8 km about 10 Ma BP, has experienced crustal uplift of around 4000–5000 m Ma^{-1} since that time.

Even these rates, though, are modest in comparison with those estimated for sections of the Southern Alps in New Zealand. This mountain range, located along the boundary of the Pacific and Indian Plates, has apparently sustained a rate of crustal uplift averaging up to 10,000 m

Ma^{-1} over the past 1 Ma. This extraordinarily high uplift rate is, as in the case of the Transverse Ranges of southern California, related to the oblique convergence occurring along the associated plate boundaries.

Schumm (1963), in an earlier review, suggested that modern rates of orogeny probably average about 7.6 m 1000 a^{-1} (7600 mm 1000 a^{-1}) and pointed out that such a rate was about eight times greater than the average maximum rate of denudation. More recent data that show an approximate equilibrium between denduation rates and crustal uplift rates have led Gilchrist and Summerfield (1991) to question that assertion.

New Zealand, as already mentioned, provides a spectacular example of a location where rates of uplift are high as a consequence of being in an environment of active plate collision and subduction. Wellman (1988) indicates that present-day uplift rates in New Zealand are generally over 200 mm 1000 a^{-1}, but that extremely high rates of 3000–17,000 mm 1000 a^{-1} occur along parts of the Southern Alps. However, uplift is not universal in New Zealand: the Central Volcanic Region of the North Island, for example, an area of back-arc spreading, is sinking at a rate of over 2000 mm 1000 a^{-1}. Figure 8.4 demonstrates the relationship between New Zealand's tectonic setting and its rates of uplift.

Rates of orogenic uplift may well vary in time. This can be illustrated by a consideration of some uplifted terraces from Japan and New Zealand of both Holocene and Late Pleistocene age (Berryman, 1987). In both areas (table 8.5), the late Holocene rates appear to be significantly faster than those for the Last Interglacial. The reasons for this difference are not clear. However, some uplift is spasmodic in that rather than being a long-continued and relatively gentle process it takes place in association with one major seismic event. Table 8.6 shows the amount of vertical displacement that has been measured for some specific earthquake events over the last century or so. It is quite possible for the equivalent of some thousands of years of 'normal' uplift to take place in just one sudden spasm.

Perhaps the largest recorded event over the last century is the Yakutat Bay earthquake, which caused over 14 m of vertical displacement in Alaska in 1899. Broadly similar amounts of displacement also took place in the 1964 Alaskan earthquake. While the amount of change can be substantial the speed of change is even more so. As Kukal (1990, p. 13) has remarked, 'As compared to other geological processes, the rate of a single seismic displacements (sic) is enormous. The actual motions in seismic shocks may, of course, be fast as shear waves (hundreds of metres or kilometres per second).'

It is possible for some individual tectonic events to affect a large area. One of the finest examples of this is provided by an event which occurred in the eastern Mediterranean about 1550 BP. It

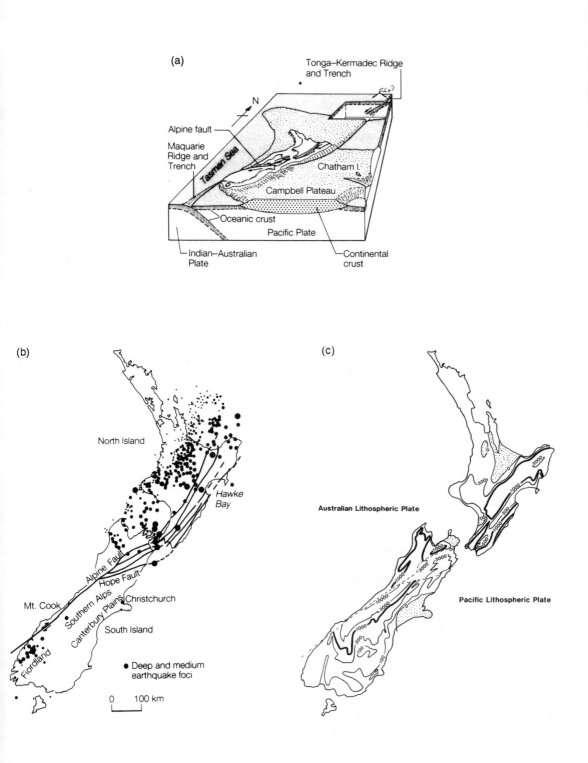

(a)

Tonga–Kermadec Ridge and Trench

N

Alpine fault

Maquarie Ridge and Trench

Tasman Sea

Chatham I.

Campbell Plateau

Oceanic crust

Pacific Plate

Indian–Australian Plate

Continental crust

(b)

North Island

Hawke Bay

Alpine Fault

Hope Fault

Mt. Cook

Southern Alps

Canterbury Plains

Christchurch

South Island

Fiordland

● Deep and medium earthquake foci

0   100 km

(c)

Australian Lithospheric Plate

Pacific Lithospheric Plate

Table 8.5 Average rates of tectonic uplift in mm 1000 a⁻¹ estimated from heights of palaeoshorelines of Holocene and Last Interglacial age in (A) Japan and (B) New Zealand

Locality	Uplift rate– Holocene	Uplift rate– last interglacial	Ratio Holocene/ last interglacial
A Japan			
1. Shirakami Mts	1300	700	1.9
2. Sado Island	1200	900	1.3
3. Nyu Mts	1000	900	1.1
4. Oiso Hills	3800	1200	3.2
5. Boso Peninsula	1500	1000	1.5
6. Muroto Peninsula	4000	1500	2.7
7. Kikai Island	1800	1500	1.2
B New Zealand			
1. Waihau Bay	1000	600	1.7
2. Te Araroa	2900	2400	1.2
3. Mahia Peninsula	3100	1300	2.4
4. Cape Kidnappers	2600	800	3.3
5. Whareama Syncline	1300	500	2.6
6. Aorangi Anticline	2700	2000	1.3
7. Conway River	2000	800	2.5
8. Westport	800	400	2.0
9. Shannon	400	200	2.0

Source: from various sources in Berryman (1987) tables 5.1 and 5.2.

Table 8.6 Rates of vertical displacement in specific earthquake events

Specific events location	Date	Displacement
Alaska	1964	10 m–15 m
Yakutat Bay (Alaska)	1899	14.3 m
New Zealand (Murchison quake)	1929	5 m
Adelaide (Australia)	1954	5–8 m
California	1872	7.01 m
Sonora (Mexico)	1887	6.10 m
Japan	1891	6.10 m
India	1897	10.67 m
California	1906	0.91 m
Formosa	1906	1.83 m
Mexico	1912	0.61 m
Nevada	1915	4.75 m

Source: from various sources in Goudie (1992) table 6.7b.

Figure 8.4
The tectonic setting of New Zealand:
(a) The position of New Zealand in relation to the Pacific Plate and the Indian–Australasian plate.
(b) The location of major faults and earthquake foci.
(c) Rates of current uplift in mm 1000 a⁻¹ (source: modified after Wellman, 1988).

affected a considerable part of Crete, Karpathos, Rhodes and the south coast of Turkey near Alanya (Kelletat, 1991). It was at its most intense in western Crete, where some 9 m of uplift appears to have taken place. Figure 8.5 shows the isobases of this event for Crete.

The changing rates of orogeny through time can be further demonstrated for Japan (see figure 8.6). This gives a picture of vertical movements in Japan averaged over a 75-year period. Figure 8.6a, clearly shows the effect of the 1946 Nanki Earthquake which followed a fifty-year period of vertical movements free of major earthquakes in Shikoku.

Over the last 40 million years, rates of uplift have varied in the Andes because of changes in the nature of plate subduction over that period. Benjamin et al. (1987), using fission track techniques, demonstrated that uplift rates increased from 100 to 200 mm 1000 a^{-1} between 20 and 40 million years ago, to as much as 700 mm 1000 a^{-1} by 3 million years ago. They believe that this trend, illustrated in figure 8.7 can be explained in terms of a decrease in the slab angle and an increase in the convergence rate of the Nazca Plate.

Rates of uplift have also changed through time in the Himalayas, where studies based on fossil floras, sediment budgets, and radiometric and fission track data show a remarkable correspondence for the late Cenozoic. They demonstrate exponentially increasing rates of uplift from the early Miocene. Brookfield (1993) believes that this early Miocene start of exponential uplift can be related to underthrusting of Indian shield crust below the Himalaya along the Main Central Thrust and to underthrusting of Asian crust below the Pamir and possibly Karakoram.

Related to temporal variability in rates of uplift, there has also been some debate in the literature as to whether rates of uplift caused by denudational isostasy are episodic rather than continuous. Gilchrist and Summerfield (1991), in reviewing the literature on this theme, suggest that the notion that isostatic uplift in response to continuous denudation over a very long-term 'cyclic timescale' (10^7 a) is episodic in nature has become firmly established. They show, however, that the notion is 'based on a fundamental misunderstanding of the principles of flexural isostasy' (p. 555) and propose that the lithosphere experiences a continuous compensation in response to denudation.

Figure 8.5
Tectonic deformation in the Eastern Mediterranean (source: Kelletat, 1991, figs 4 and 5):
(a) Areas (stippled) that had significant uplift about 1550 BP.
(b) Isobase map of western Crete showing differential uplift during the 1550 BP neotectonic event.

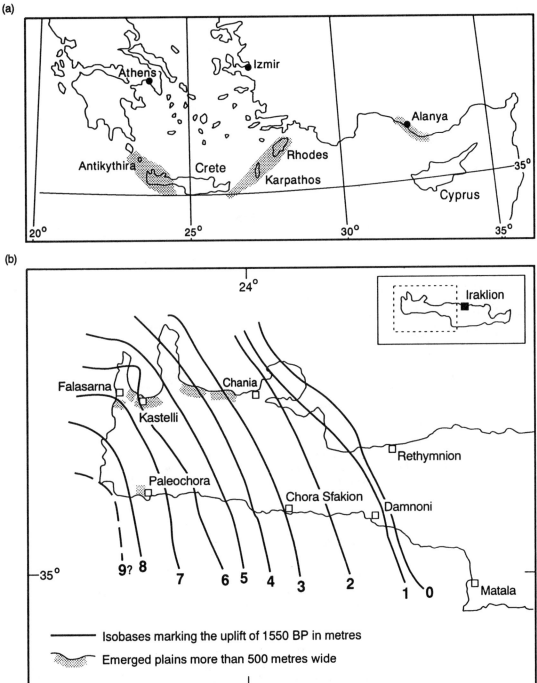

(a)

Athens

Izmir

Alanya

Rhodes

Antikythira

Crete

Karpathos

Cyprus

35°

20°　　25°　　30°　　35°

(b)

24°

Iraklion

Falasarna

Chania

Kastelli

Rethymnion

Paleochora

Chora Sfakion

Damnoni

35°

9? 8　7　6 5　4　3　2　1 0　Matala

——— Isobases marking the uplift of 1550 BP in metres

～～～ Emerged plains more than 500 metres wide

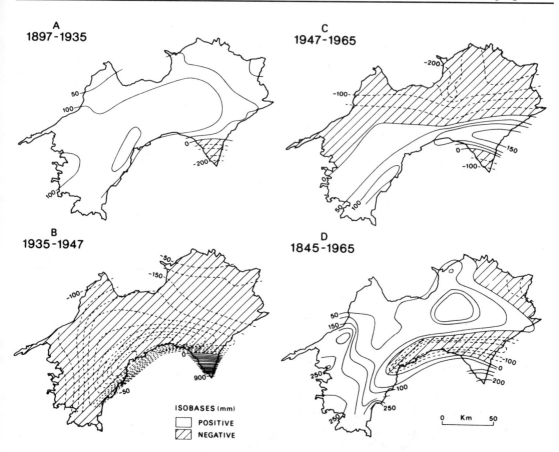

Figure 8.6
Vertical displacement of Shikoku, Japan, during three time periods (A–C) and for the total period 1845–1965 (D). Those values partly based on precise levelling are shown by 50 mm isobases; the effect of the 1946 earthquake is particularly apparent (source: Yoshikawa in Chorley, Schumm and Sugden, 1984, fig. 3.23).

Salt tectonics

Evaporites, such as halite (sodium chloride) are widespread both in the geological record and in terms of their distribution. Salts flow readily under burial conditions, and this can lead to the development of various geological structures including domes and diapirs. Their structures tend to evolve from concondant, low-amplitude intrusions, through to higher amplitude intrusions, and thence to extrusions. Once extrusion takes place the rate of growth may slow because of solutional losses and the flow of salt glaciers (*namakiers*) (Goudie, 1989).

The growth of domes and piercing diapirs causes crustal uplift, which may be at a sufficient rate to cause distortion of drainage

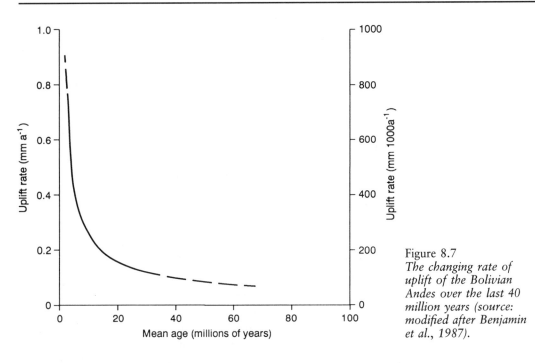

Figure 8.7
The changing rate of uplift of the Bolivian Andes over the last 40 million years (source: modified after Benjamin et al., 1987).

systems and the development of clearly defined surface forms. There are, however, relatively few data on the rates at which doming of the ground surface occurs. For the Sigsbee Knolls in the Gulf of Mexioo, Ewing and Ewing (1962) give a rate of about 1000 mm 1000 a^{-1}, while for East Texas, Seni and Jackson (1983) give net rates of between 26 and 220 mm 1000 a^{-1}. Talbot and Jarvis (1984, p. 531) believe that a 'commonly accepted figure for the rate of rise of the top surface of salt diapirs averaged over geological periods is about 2000 to 4000 mm 1000 a^{-1} although local spines have been known to rise 15,000 mm 1000 a^{-1} for brief periods'. These rates are not as high as some of those recorded from areas of 'normal' active orogenic uplift (see table 8.4 for comparison) but are of sufficient magnitude to be geomorphologically significant.

Epeirogenic uplift

Epeirogeny is the uplift of what are usually large areas of the earth's surface without significant folding or fracture. It is essentially associated with plate interiors rather than colliding plate margins. There is a range of processes that can account for plate centre movements including broad scale doming and rifting associated with hotspots, and subsistence associated with lithospheric cooling. Summerfield (1991a, chapter 4) provides a review of the mechanisms involved.

Although data are still relatively sparse on long-term rates of uplift for zones of epeirogeny, one of the most fascinating aspects of this issue is the fact that rates of uplift can be appreciable in such situations. Summerfield (1991a, p. 378) suggests that their rates of uplift, averaged over millions of years, appear to lie in the range of 10 to 200 m Ma^{-1}. These rates may be less than those for orogenic uplift, but they are still substantial.

Plainly, rates of orogenesis are intimately related to their structural or plate tectonic setting. Fairbridge (1981) gives the following summary data (in mm 1000 a^{-1}) on recent vertical crustal movement rates for different geotectonic settings:

Shields and platforms	1
Cenozoic orogenic belts	20,000
Older Phanerozoic orogenic belts	5,000
Intra-orogenic Basin-Range belts (with regional block faulting)	10,000

If one takes the most active of these categories, the Cenozoic orogenic belts, one can also see whether it is possible to see rates of uplift as a function of the degree of plate interaction. Yonekura (1987, cited in Muhs et al. 1990) suggested that in the circum-Pacific area rates should be least in intraplate locations (oceanic islands or atolls), low in ocean to ocean subduction zones (island arcs), moderate to high in ocean to continent subduction zones (continental arcs), and very high in accretion or collision zones where relatively buoyant crust (such as aseismic ridges, seamounts or other topographic highs) on the lower plate collides with a buoyant continental upper plate. However, after collecting data from a large number of sites, they found little sound evidence, determined on the basis of the heights of marine terraces from the Last Interglacial, to support this contention. They found (figure 8.8) that the very highest values for each category followed the Yonekura model, but that if these were discounted then rates were remarkably similar, clustering between 0 and 1250 mm 1000 a^{-1}.

The details of tectonic setting are also important in explaining the highly variable rates of uplift and subsidence that are made evident from a study of tide gauge trends in Japan. Aubrey and Emery (1986) have analysed these trends and show that on average Hokkaido has a subsidence rate of 2900 mm 1000 a^{-1}, Honshu of 4000 mm 1000 a^{-1} (5100 mm 1000 a^{-1} for East and South Honshu and 2200 mm 1000 a^{-1} for West Honshu) and Shikoku of 3200 mm 1000 a^{-1}. By contrast Kyushu appears to be emerging at a rate of 600 mm 1000 a^{-1}. The explanation proffered for this is that subduction along the Kuril, Japan, Bonin, Nankai and Ryukyu-Kyushu trenches draws the over-riding continental plate downward, resulting in continental tilt towards the trench. Along the Kuril and Japan trenches this results in a subsidence of eastern

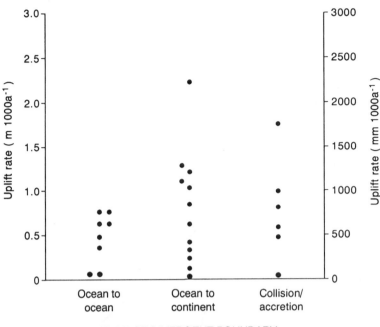

Figure 8.8
Rates of long-term vertical uplift for three different plate tectonic settings (source: modified after Muhs et al., 1990).

Hokkaido and Honshu. The convergence pattern along the south coasts of Japan is more complex due to the geometry of the Bonin, Nankai and Ryukyu-Kyushu trenches. Relative sea-level trends reflect this complex underthrusting through two distinct blocks; north of the Median Tectonic line along Kyushi, Shikoku and south-west Honshu, the first block tilts towards the south; whereas south of the Median Tectonic Line the second block is being raised at the north and lowered at the south.

On passive continental margins there are contrasting patterns of uplift and subsidence (Cronin, 1981). Long-term lithospheric flexural upwarping takes place on the continental margin, and subsidence occurs offshore, partly because of sediment loading. Along the eastern seaboard of the USA, Cronin found that the rates of upwarping averaged about 10–30 mm 1000 a^{-1}, and that the rates of subsidence offshore averaged 20–40 mm 1000 a^{-1}. These rates were based on the height relationships of dated marine terraces. Kukal (1990) provides a more general survey of the rates at which subsidence may occur in certain particular tectonic settings, such as continental margins. These are geological units where large thicknesses of sediment and relatively rapid subsidence are typical. In particular, passive continental margins are areas of long-term subsidence, and empirical curves of the mean subsidence rate over extended periods can be obtained using boreholes and geophysical methods. Emery and Uchupi (1972), for example, calculated rates for the continental margin of the United States back to

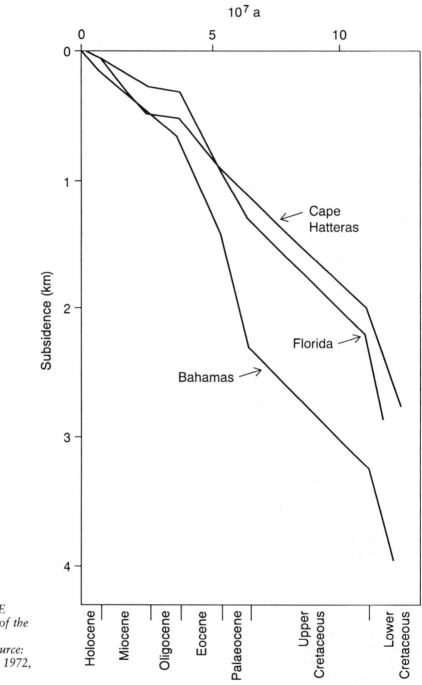

Figure 8.9
Subsidence at the SE continental margin of the USA expressed by empirical curves (source: Emery and Uchupi, 1972, with modifications).

Table 8.7 Rates of continental subsidence

	mm 1000 a⁻¹
Danube Valley (Czechoslovakia)	800–1000
Lake District	800
Kamchatka	4000
Basins of South California, USA	1200–2000
Caspian and Black Sea Basins	300
East Germany	2000–5000
Rhine Graben	500

Source: Kukal (1990) p. 34.

the Cretaceous (figure 8.9). The shelf appears to be subsiding fastest in the Gulf of Mexico (at $c.400$ mm 1000 a^{-1}) and slowest in Yucatan (at $c.1$ mm 1000 a^{-1}), but rates for the eastern seaboard appear to average 20–30 mm 1000 a^{-1}.

There are also areas of continental subsidence away from continental margins. Subsidence may take place in basins and rifts at an appreciable rate. Kukal (1990, p. 34) cites rates for various areas, and these are summarized in table 8.7. Rates typically range from 300 to 5000 mm 1000 a^{-1}.

Island subsidence

Crucial to Darwin's model of atoll evolution is the idea that subsidence has occurred, and the presence of guyots and seamounts in the Pacific Ocean attests to the fact that such subsidence has been a reality over wide areas. In addition there are tide gauge records from the Hawaiian Ridge that demonstrate ongoing subsidence rates at the present day (1500 mm 1000 a^{-1} for Oahu and 3500 mm 1000 a^{-1} for Hawaii). Moreover, the occurrence of Holocene and Pleistocene submerged terraces and drowned reefs on the flanks of these two islands demonstrates the reality of this process. The thick coral accumulations that are superimposed on basaltic platforms that were once at sea-level, indicate that subsidence has continued on time-scales of tens of millions of years. The coral cap at Eniwetok, which dates back to the early Eocene ($c.60$ Ma ago) is 1400 m thick, and that at Bikini (of Miocene age) some 1300 m thick. Differences in the depths of waveworn platforms along a volcanic chain, if the chronology of the seamount formation can be established by potassium-argon and other dating techniques, provide estimates of subsidence rates. In the case of the Tasman Sea chains off Australia, rates of subsidence appear to have been of the order of 28 m Ma^{-1} (28 mm 1000 a^{-1}) (figure 8.10).

The causes of this subsidence are a matter of some debate (see Lambeck, 1988, pp. 506–9). Some of it may be caused by the loading of volcanic material onto the crust, but some may be due to a gradual contraction of the sea floor as the ocean lithosphere

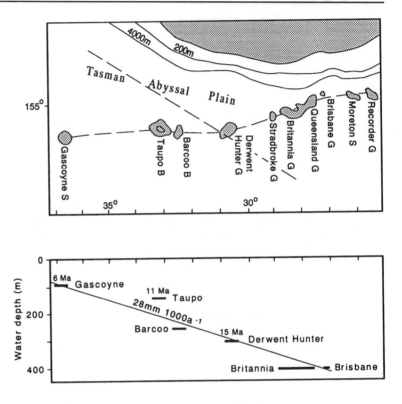

Figure 8.10
The Gascoyne–Recorder seamount chain in the Tasman Sea to the east of Australia. (S = seamount, B = Bank, G = Guyot). The spreading axis of the Tasman Sea runs approximately north–south. On the right the depths of the seamount platforms are plotted as a function of latitude. Their potassium-argon dates (in millions of years) were determined by I. McDougall, and the average subsidence rate for the chain is shown by the continuous line and is in mm 1000 a^{-1} (source: modified after Lambeck, 1988, fig. 10.15).

moves away from either the ridge or the hotspot that led to the initial formation of the island volcanoes. The whole question of earth movements associated with the building of volcanic edifices has recently been reviewed by McGuire and Saunders (1993).

Island emergence

Not all Pacific ocean islands show subsidence. On others there are elevated marine terraces, notches and coral reefs. Examples include Makatea, Mangaia, Nauru and Ocean Island in the south Pacific. Terraces attributed to the Last Interglacial, around 120,000 years ago, may be elevated by 20 metres or so, indicating an uplift rate of around 200 mm 1000 a^{-1}.

The uplifts of some of the old atolls, as seen in the elevated reef platforms of both the Cook and Society island groups, appear to be a function of distance from the major centres of recent volcanism of Rarotonga and Tahiti respectively. It has been proposed by McNutt and Menard (1978) that the explanation for this pattern is to be found in terms of the regional isostatic response to volcanic

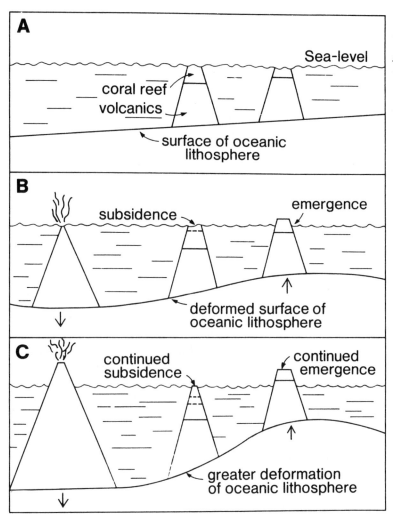

Figure 8.11
A model (based on McNutt and Menard 1978) of apparent sea-level change brought about in the vicinity of coral reefs by volcanic loading on to the elastic lithosphere: (A) At around one million years ago atolls developed on two extinct volcanoes are in equilibrium with sea-level on an oceanic lithosphere surface that is slowly sinking as it cools (right to left); (B) By c.0.3 million years ago, active volcanism loads the lithosphere and causes subsidence in a near-field 'moat' and emergence at a greater distance (typically c.200 km). In the moat region new reef formation is induced by the subsidence and reefs become emerged at a greater distance; (C) At the present time, continued volcanic activity produces still more volcanic material so that subsidence and uplift continue (source: Goudie, 1992, fig. 6.4).

loading shown in figure 8.11, but this is not a universally accepted model, and it is possible (Lambeck, 1988, p. 512) that the uplift results from a broad doming of the sea floor associated with the mantle hotspots that led to the formation of the recent volcanoes.

Glacio-isostasy

As we have just seen in the context of the possible role of volcanic loading, the earth's crust responds when a load is either applied to or removed from it. The area of crust immediately under a load is depressed, while at some distance away there is a compensatory uplift of the crust. When a load is removed from an area, uplift

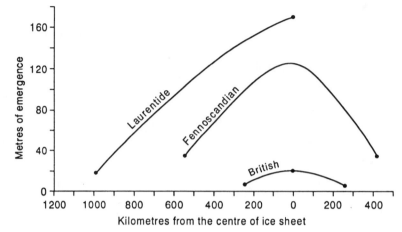

Figure 8.12
Cross section through the
Laurentide, Fennoscandian
and British ice sheets
showing the amounts of
isostatic recovery in the
last 7000 years (source:
modified from Andrews,
1975).

occurs, but at some distance away compensatory subsidence may
occur.

One particular type of isostasy, glacio-isostasy, results from the
growth and decay of ice masses on the earth's crust. It was a very
important process during the multiple glacial–interglacial cycles of
the Pleistocene, and has greatly influenced the sea-level history of
many parts of the world during the Holocene.

The degree of isostatic change, and the rate at which it occurred
is related to the differing volumes of the various ice caps (figure
8.12). This is illustrated when one looks at the amount of isostatic
recovery that has taken place in the Holocene in North America,
Fennoscandia and the British Isles. It has been greatest in the area
vacated by the Laurentide Ice Cap, the largest of the three ice caps
under consideration, and least in the area vacated by the British
Ice Cap, which was the smallest of the three.

The rate of isostatic uplift in Norway has been described by
Hafsten (1983), and his data are summarized in table 8.8 and
figure 8.13. The pattern of uplift rates declines westwards, which
broadly reflects the decreasing thickness of the Scandinavian ice
sheet as one moves away from its centre near the Gulf of Bothnia.
Also notable is the way in which the rate of isostatic adjustment
has changed through time. The most rapid uplift took place in the
early Holocene (i.e. in the Preboreal and Boreal), with about two-
thirds of the total Holocene adjustment taking place prior to 8000
BP.

Moving to Sweden and Finland (Eronen, 1983) one can see that
the fastest rates of current adjustment occur at the head of the
Gulf of Bothnia (9000 mm 1000 a^{-1}), and decline progressively
towards the southern Baltic. The rapid rates of rise in the Holocene
mean that there has been as much as 290 m of displacement in
Sweden and 220 m in Finland.

Table 8.8   Holocene shoreline displacement resulting from glacio-isostatic adjustment in five areas of eastern Norway

	Preboreal and Boreal 10,000–8000 BP			Atlantic 8000–5000 BP			Subboreal and Subatlantic 5000 BP			
Region	Highest marine limit (m)	Total shore displacement (m)	Mean shore displacement per 100 years (m)	Total Holocene shore displacement (m)	Total shore displ. (m)	Mean shore displacement per 100 yrs (m)	Total Holocene shore displ. (m)	Total shore displ. (m)	Mean shore displacement per 100 yrs (m)	Total Holocene shore displ. (m)
Oslo area	220	150	7.5	68	45	1.5	20	25	0.5	11
Ski area	215	140	7.0	6.5	38	1.3	18	38	0.8	18
S. Østfold	185	130	6.5	70	35	1.2	19	20	0.4	11
Frosta	175	115	5.8	66	30	1.0	17	28	0.5	16
S. Vestfold	155	105	5.3	68	25	0.8	16	25	0.5	16

Source: from Hafsten (1983) table 7.

In Greenland very rapid rates of adjustment to ice load removal are evident in the early Holocene, and Pirazzoli (1991) reports on the available data. At Mesters Vig early Holocene emergence was at a rate of 90,000 mm 1000 a^{-1}, decreased exponentially to about 6000 mm 1000 a^{-1} for the interval 9000 to 6000 years BP, and has remained perhaps as low as 700 mm 1000 a^{-1} since 6000 BP. On the west coast, at Søndre Strømfjord, the initial rate of uplift was at about 105,000 mm 1000 a^{-1}.

The changes that have taken place in Arctic Canada have been the subject of a detailed monograph by Andrews (1970). Figure 8.14 shows the average rate of uplift for Northern and Eastern Canada for the whole Holocene, while figure 8.15 shows the present rate of uplift. At the south-east end of Hudson Bay the average rate has approached 30 m 1000 a^{-1} (35,000 mm 1000 a^{-1}), and at present reaches 13,000 mm 1000 a^{-1}. In the early Holocene rates were very much larger than either the average or present figures, for as in Norway and Greenland, the great bulk of the uplift took place in the first few thousands of years following deglacierization (table 8.9). Note also how closely the area of maximum rate of isostatic uplift corresponds to the area of maximum Laurentide ice thickness.

Crustal depression caused by deltaic loading

One particular type of isostasy is that caused by the loading of sediment onto the earth's crust. A pertinent example of this phenomenon is the way in which crustal subsidence is caused by deltaic sedimentation. Some deltas show dated Holocene sediments at appreciable depths, and in such cases it is possible to calculate the rate at which loading-induced subsidence has occurred. Fairbridge (1983, p. 14) provides an analysis of the situation with respect to the Mississippi Delta in the USA:

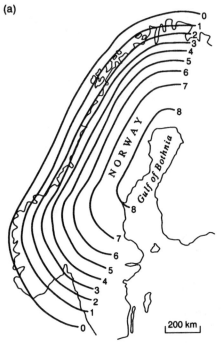

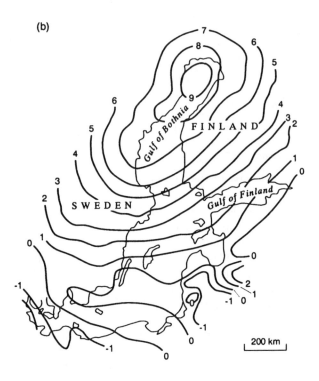

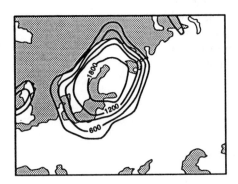

On the top of the delta, in a roughly central position, the floor of the Holocene is found commonly at a depth of about 200 m. At 10,000 BP the eustatic level was about 35 m, so that an anomaly of up to 165 m must be explained by tectonic subsidence and sedimentary compaction. In a mixed sandy, clayey and peaty sequence it is difficult to calculate the latter with precision, but the bulk of the compaction is accommodated in the top 10 m, so that in a 165 m Holocene sequence, something like 15 m may be explained that way, leaving 150 m as the isostatic-tectonic component, or 15.0 mm a^{-1}.

Likewise, Stanley and Chen (1993) have calculated Holocene subsidence rates for the Yangtze river delta of China, and at the seaward end find rates of the order of 1.6–4.4 mm a^{-1} (16,000–44,000 mm 1000 a^{-1}).

Humanly-induced ground subsidence

The subsidence of the land surface is a widespread phenomenon, and can result from a range of different mechanisms, the importance and rate of operation of which can be influenced by human activities: the withdrawal of groundwater, oil and gas; the extraction of coal, salt, sulphur and other solids through mining; the hydrocompaction of sediments; the oxidation and shrinkage of organic deposits such as peats and humus rich soils; the melting of permafrost; and the catastrophic development of sink-holes in karstic terrain.

The most spectacular of these mechanisms is the withdrawal of fluids. Subsidence produced by oil abstraction has been a problem in many parts of the world (table 8.10a), and the classic area is Los Angeles in California where over 9 m of subsidence has occurred as a result of the exploitation of the Wilmington oilfield between 1928 and 1971.

More widespread is ground subsidence produced by groundwater abstraction for industrial, domestic and agricultural purposes. Data are presented in table 8.10b. Particularly severe subsidence, amounting to over 8 m in the Central Valley of California, occurs in arid areas, where groundwater resources are mined at rates that are sometimes much greater than they can be replenished. However, it is also an increasing problem in some major metropolitan areas, including Bangkok in Thailand, and Tokyo in Japan, and as

Figure 8.13
The glacio-isostatic situation in Scandinavia:
(a) Isobases for present-day land uplift in Norway in mm per year
    (source: Hafsten, 1983, fig. 5).
(b) Isobases of present-day land uplift in Sweden and Finland in mm
    per year (source: Eronen, 1983, fig. 1).
(c) Ice thickness at 18,000 years ago (contours in metres).

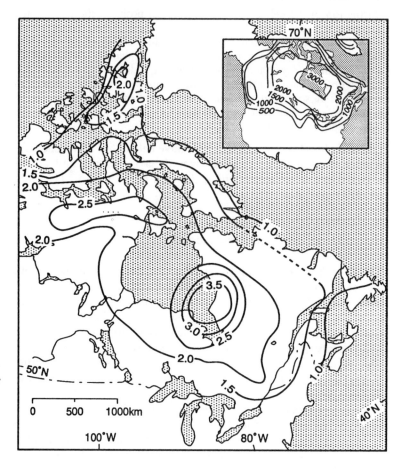

Figure 8.14
Average rate of Holocene uplift in metres per 100 years for northern and eastern Canada. The inset shows the thickness of the Laurentide ice sheet at 18,000 BP (contours in metres) (source: Andrews, 1970, fig. 2.7, with modifications).

a result the risk of flooding has been greatly accelerated. Some areas have subsided below sea level.

The data for Mexico City (figure 8.16), where over 6 m of subsidence has occurred, show the relationship between population growth, water consumption, groundwater pumping and subsidence (Rivera et al., 1991).

Land drainage can promote subsidence by accelerating the decay of organic soils. The lowering of the water table makes the peat and humus more susceptible to oxidation and deflation, so that its volume decreases. One of the longest records of this process is provided by the Holme Fen Post in the English Fenland (figure 8.17). Approximately 3.8 m of subsidence occurred between 1848 and 1957, with the fastest rate occurring soon after drainage had been initiated. The present rate averages about 1.4 cm a^{-1} (14,000 mm 1000 a^{-1}) (Richardson and Smith, 1977).

A further type of accelerated subsidence, sometimes associated

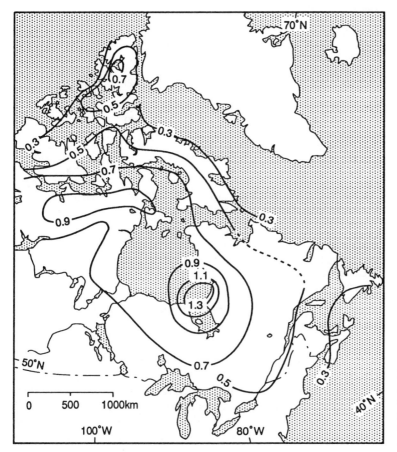

Figure 8.15
Present rate of uplift in metres per 100 years for northern and eastern Canada (source: Andrews, 1970, fig. 8.10).

Table 8.9 Percentage of uplift for thousand-year intervals during glacio-isostatic compensation

Time since uplift began	Mean %	Standard deviation %
0	0	–
1	33.0	4.4
2	56.0	5.7
3	70.4	5.8
4	80.2	5.4
5	86.5	4.5
6	91.1	3.7
7	93.9	2.9
8	94.6	3.0
9	97.4	1.9
10	100	–

Source: from Andrews (1970) table III-3.

Table 8.10 Ground subsidence

Location	*Amount (m)*		*Rate (mm 1000 a^{-1})*
(a) Ground subsidence produced by oil and gas abstraction			
Azerbaydzhan, CIS	2.5	(1912–62)	50,000
Atravopol, CIS	1.5	(1956–62)	125,000
Wilmington, USA	9.3	(1928–71)	216,000
Inglewood, USA	2.9	(1917–63)	63,000
Groningen, Netherlands	0.18	(1917–63)	9,000
(b) Ground subsidence produced by ground-water abstraction			
London, England	0.06–0.08	(1865–1931)	910–1210
Savanna, Georgia	0.1	(1918–55)	2,600
Mexico City	7.5	–	250,000–300,000
Houston, Galverston, Taxas	1.52	(1943–64)	60,000–76,000
Central Valley, California	8.53	–	–
Tokyo, Japan	4	(1892–1972)	500,000
Osaka, Japan	> 2.8	(1935–72)	76,000
Niigata	> 1.5	–	–
Pecos, Texas	0.2	(1935–66)	6,500
South-central Arizona	2.9	(1934–77)	96,000
Bangkok, Thailand	0.5	–	100,000

Source: from various sources in Goudie (1993) table 6.4, and Doornhof (1992).

with increased and more intense seismic activity, results from the effects on the earth's crust of large masses of water impounded as reservoirs behind dams. This process, which is called hydro-isostasy, can lead to measurable amounts of crustal downwarping, typically in the range of 5–20 cm in a decade after filling of the reservoir. Table 8.11 provides some data for some major dam schemes.

The formation of volcanoes

Finally, let us look at one further consequence of endogenic activity – the formation of volcanoes. These are the most obvious manifestations of crustal activity, and there are at present some 600 active volcanoes on the continents or exposed above the sea as islands. In addition there are huge numbers of submarine volcanoes – some 50,000 or more on the floor of the Pacific alone (Summerfield, 1991a, p. 109). Individual volcanic events can be spectacular and produce large amounts of volcanic material in a short period of time. However, the average amount of material produced over extended periods of time is rather less spectacular. The discharge from volcanoes forming island arcs and continental margin orogens is only about 1 km^3 a^{-1}. The amount of material produced by hotspot related volcanoes is rather greater. The island of Hawaii alone seems to have a long-term rate of construction of over 0.4 km^3 a^{-1}, while Iceland has sustained a rate of

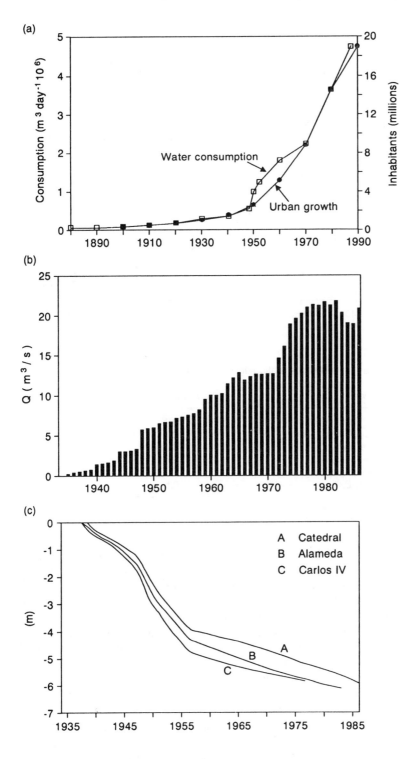

Figure 8.16
Some changes in Mexico City (source: Rivera et al., 1991, figs 1, 2 and 3):
(a) Population and water consumption.
(b) Pumping rate of groundwater.
(c) Rate of subsidence at three sites in the downtown area.

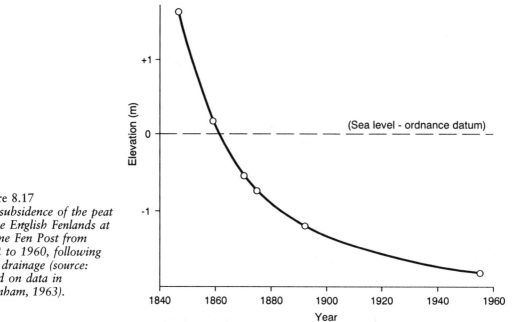

Figure 8.17
The subsidence of the peat in the English Fenlands at Holme Fen Post from 1842 to 1960, following land drainage (source: based on data in Fillenham, 1963).

Table 8.11   Hydro-isostatic subsidence caused by the creation of reservoirs

Reservoir	Maximum downwarping (cm)	Period
Lake Mead, USA	20.1	1950–63
Kariba, Central Africa	12.7	1959–68
Koyna, India	8–14	1962–8
Bratsk, Siberia	5.6	1961–6
Krasnoyarsk, Siberia	3	1967–71
Plyava, Baltic	0.5	1965–70

Source: from data in Nikonov (1977).

about 0.13 km^3 a^{-1} during historic time (Summerfield, 1991a, p. 110). The fast rate of deposition in association with hotspots means that large volcanoes can form in rapid fashion. In the case of Hawaii about 600,000 years seems to be needed to build them up from the sea floor to their present eminence (Moore and Clague, 1992). However, the volcanic activity that makes the Hawaiian islands is bound to be relatively short lived as they will be carried away from their magma source by plate motion (see p. 225). By contrast, Iceland, built on a mid-ocean ridge, has not moved very significantly with respect to its hotspot since the opening of the North Atlantic by sea-floor spreading over 50 million years ago. It has succeeded in accumulating a 10 km thick pile of volcanic material, which has five times the volume of Hawaii, in spite of the long-term rate of volcanic activity being much less.

Conclusions

The importance of tectonic forms was well expressed by that great exponent of the importance of tectonic style for landform evolution, W. Penck (1953, p. 11):

... the modelling of the earth's surface is determined by the ratio of the intensity of the endogenetic to that of the exogenic displacement of material.

A brief survey of the earth's surface shows that the ratio very often changes, or has changed, to the prejudice of the exogenic forces; the accumulation of a volcanic cone is possible only because it takes place more rapidly than the removal by denudation of the accumulated material. Faults can become visible as unlevelled fault scarps, for instance in the zone of the rift valleys of East and Central Africa, only when the formation of faults takes place more rapidly than levelling by denudation. Generally speaking, the origin of any outstanding elevation, any mountain mass, is bound up with the assumption that mountain building is more successful, i.e. works more rapidly, than denudation. Thus the varied altitudinal form of the land shows that in many cases the work of denudation is lagging behind the endogeneic displacement of material, here more, there less, or has done so in the past. The one consistent feature, however, common to every region, is that the activity of exogenetic happenings is subordinate to that of endogenetic processes.

However, endogeneic or tectonic processes have gained a greater prominence in geomorphological thought in the last few decades because of the arising and development of the plate tectonic paradigm in the 1960s. Many areas of the Earth's crust that were once thought to be relatively stable or quiescent tectonically speaking can now be seen as having a complex tectonic history and to be more active at the present time than was previously thought. Former concepts of a few widely-spaced orogenic spasms spread through hundreds of millions of years of geological time have now been replaced by a vision of more continuous and on-going change. This revolutionary modification in views on tectonic history and current rates of tectonic change has caused a major reassessment, for example, of the long-term landscape evolution and denudation chronology of the British Isles (Goudie 1991). Most importantly of all, we now know that rates of sea-floor spreading, plate collision and plate subduction are such that the world's oceans are all extraordinarily young when considered in the light of the total age of the Earth. Equally, the positions of the continents have shifted rapidly even in the span of the Tertiary, with profound climatological and geomorphological consequences. Major mountain chains have also been uplifted, often in just a few millions of years, and large volcanic mountains (e.g. in Hawaii) can develop in under a million years. Plate margins can show rapid rates of both horizontal and vertical movement, though, as we have demonstrated from New Zealand and Japan, the rates can be highly variable both in space and time, with episodic extreme events creating substantial change at specific places at specific points in time.

Even away from colliding plate margins broad-scale epeirogenic processes may operate at measurable and significant rates, while on passive continental margins (e.g. the eastern USA, western India, and southern Africa) patterns of flexural warping have been identified, with uplift of the continental margin and subsidence offshore. Even in the middle of oceanic plates, volcanic activity and crustal subsidence may occur, understanding of which is crucial to an appreciation of coral atoll formation.

Also highly significant for landform evolution is the role of various types of isostasy caused by loading of the earth's crust with substances such as ice, water, sediment, or volcanic lava. Especially notable has been the rapidity with which the Earth's crust over Laurentia and Scandinavia has responded to the rapid glacial and interglacial cycles of the Quaternary.

Finally, at the present day, humans are starting to influence endogenic processes, most especially by the removal or application of fluids or solids to the Earth's surface. In some areas of groundwater or hydrocarbon removal, subsidence can be accelerated to as much as $c.10$ m per century.

References

Aaby, B. and Tauber, H. 1974: Rates of peat formation in relation to degree of humidification and local environment as shown by studies of a raised bog in Denmark. *Boreas*, 4, 1–17.

Abrahams, A. D. and Marston, R. A. 1993: Drainage basin sediment budgets: an introduction. *Physical Geography*, 14, 221–4.

Abrahams, A. D., Howard, A. D. and Parsons, A. J. 1994: Rock-mantled slopes. In A. D. Abraham and A. J. Parsons (eds), *Geomorphology of Desert Environments*, London: Chapman and Hall, 173–212.

Ackroyd, P. 1986: Debris transport by avalanche, Torlesse Range, New Zealand. *Zeitschrift für Geomorphologie*, 30, 1–14.

Adams, J. 1980: High sediment yields from major rivers of the western southern Alps, New Zealand. *Nature*, 287, 88–9.

Adepegba, D. and Adegoke, E. A. 1974: A study of the compressive strength and stabilising chemicals of termite mounds in Nigeria. *Soil Science*, 117, 175–9.

Adey, W. H. 1978: Coral reef morphogenesis: a multidimensional model. *Science*, 202, 831–7.

Ahnert, F. 1970: Functional relationships between denudation, relief and uplift in large mid-latitude drainage basins. *American Journal of Science*, 268, 243–63.

Åkerman, H. J. 1993: Solifluction and creep rates 1972–1991, Kapp Linné, west Spitsbergen. In B. Frenzel, J. A. Matthews, and S. B. Gläser (eds), *Solifluction and climatic variation in the Holocene*. Stuttgart: Gustav Fischer, 225–249.

Alexander, L. T. and Cady, J. G. 1962: Genesis and hardening of laterite in soils. *Technical Bulletin United States Department of Agriculture*, 1282.

Allegré, C. 1988: *The behaviour of the Earth*. Cambridge, Mass.: Harvard University Press.

Allen, J. R. L. and Rae, J. E. 1988: Vertical salt-marsh accretion since the Roman period in the Severn Estuary, southwest Britain. *Marine Geology*, 83, 225–35.

Amit, R., Gerson, R. and Yaalon, D. H. 1993: Stages and rate of the gravel shattering process by salts in desert reg soils. *Geoderma*, 5, 295–324.

Anderson, E. W. and Cox, N. J. 1984: The relationship between soil creep rate and certain controlling variables in a catchment in upper Weardale, northern England. In T. P. Burt and D. E. Walling (eds),

Catchment experiments in fluvial geomorphology, Norwich: Geobooks, 419–30.

Anderson, J. G. 1906: Solifluction, a component of subaerial denudation. *Journal of Geology* 14, 91–112.

Anderson, R. S., Borns, H. W., Smith, D. C. and Race, C. 1992: Implications of rapid sediment accumulation in a small New England salt marsh. *Canadian Journal of Earth Science*, 29, 2013–17.

André, M-F. 1988: Vitesses d'accumulation des débris rocheux au pied des parois supraglaciaires du nord-ouest du Spitsberg. *Zeitschrift für Geomorphologie*, 32(3), 351–73.

André, M-F. 1990: Geomorphic impact of spring avalanches in Northwest Spitsbergen. *Permafrost and Periglacial Processes*, 1, 97–110.

Andrews, J. T. 1970: A geomorphological study of post-glacial uplift with particular reference to Arctic Canada. *Institute of British Geographers Special Publication 2*.

Andrews, J. T. 1975: *Glacial Systems – an approach to glaciers and their environments*. North Scituate: Duxbury Press.

Andrews, J. T. and Mahaffy, M. A. W. 1976: Growth rate of the Laurentide Ice Sheet and sea level lowering (with emphasis on the 115,000 BP sea level low). *Quaternary Research*, 6, 167–83.

Ashmore, P. 1993: Contemporary erosion of the Canadian landscape. *Progress in Physical Geography*, 17, 190–204.

Atkinson, T. C., Harmon, R. S., Smart, P. L. and Waltham, A. C. 1978: Palaeoclimatic and geomorphic implications of $s^{230}Th/^{234}U$ dates on speleothems from Britain. *Nature*, 272, 24–8.

Attewell, P. B. and Taylor, D. 1988: Time dependent atmospheric degradation of building stone in a polluting environment. In G. Marinos and G. Koukis (eds), *Engineering Geology of Ancient Works, Monuments and Historical Sites*, Rotterdam: Balkema.

Aubrey, D. G. and Emery, K. O. 1986: Relative sea levels of Japan from tide-gauge records. *Bulletin Geological Society of America*, 97, 194–205.

Babbage, C. 1847: Observations on the temple of Serapis, at Pozzuoli, near Naples. *Quarterly Journal of Geological Society, London*, 3, 186–217.

Bagine, R. K. N. 1984: Soil translocation by termites of the genus *Odontotermes* (Holmgren) (Isoptera: Macrotermininae) in an arid area of northern Kenya. *Oecologia*, 64, 263–6.

Bagnold, R. A. 1941: *The physics of blown sand and desert dunes*. London: Methuen.

Bain, D. C., Mellor, A., Wilson, M. J. and Duthie, D. M. L. 1990: Weathering in Scottish and Norwegian catchments. In B. J. Mason (ed.), *The surface waters acidification programme*, Cambridge: Cambridge University Press, 223–33.

Ball, J. 1927: Problems of the Libyan Desert. *Geographical Journal*, 70, 21–38, 105–28, 209–24.

Ballantyne, C. K. and Kirkbride, M. P. 1987: Rockfall activity in upland Britain during the Loch Lomond stadial. *Geographical Journal*, 153, 86–92.

Barnard, 1973: Duinformasies in die centrale Namib. *Tegnikon*, 1–13.

Barton, D. C. 1916: Notes on the disintegration of granite in Egypt. *Journal of Geology*, 24, 382–93.

Bell, M. and Laine, E. P. 1985: Erosion of the Laurentide region of North America by glacial and glaciofluvial processes. *Quaternary Research*, 23, 154–74.

Bell, M. and Walker, M. J. C. 1992: *Late Quaternary Environmental Change*, Harlow: Longman Scientific.

Benedict, J. 1993: Influence of snow upon rates of granodiorite weathering, Colorado Front Range, USA. *Boreas*, 22, 89–92.

Benjamin, M. T., Johnson, N. M. and Naeser, C. W. 1987: Recent rapid uplift in the Bolivian Andes: evidence from fission track dating. *Geology*, 15, 680–3.

Bennett, H. H. 1939: *Soil Conservation*. New York: McGraw Hill.

Berner, R. A. and Rye, D. M. 1992: Calculation of the Phanerozoic strontium isotope record of the oceans from a carbon cycle model. *American Journal of Science*, 292, 136–48.

Berryman, K. 1987: Tectonic processes and their impact on the recording of relative sea-level changes. In R. J. N. Devoy (ed.), *Sea Surface Studies a global view*, London: Croom Helm, 127–61.

Berthois, L. and Portier, J. 1957: Recherches expérimentales sur le façonnement des grains de sable quartzeux. *Comptes Rendus Académie des Sciences*, 245, 1152–4.

Bilham, R. and Simpson, D. 1984: Indo-Asian convergence, and the 1913 survey line connecting the Indian and Russian triangulation surveys. In *The International Karakoram project*, K. J. Miller (ed.), vol. 1., 160–70, Cambridge: Cambridge University Press.

Bird, E. C. F. 1985: *Coastal change: a global review*. Chichester: Wiley.

Bishop, P. 1985: Southeast Australia late Mesozoic and Cenozoic denudation rates: a test for late Tertiary increases in continental denudation. *Geology*, 13, 479–82.

Black, T. A. and Montgomery, D. R. 1991: Sediment transport by burrowing mammals, Marin County, California. *Earth Surface Processes and Landforms*, 16, 163–72.

Blackburn, W. H., Knight, R. W. and Schuster, J. L. 1983: Saltcedar influence on sedimentation in the Brazos River. *Journal of Soil and Water Conservation*, 37, 298–301.

Blackwelder, E. 1933: The insolation hypothesis of rock weathering. *American Journal of Science*, 26, 97–113.

Blong, R. J. 1984: *Volcanic hazards*. Sydney: Academic Press.

Bloom, A. L. 1978: *Geomorphology: A systematic analysis of late Cenozoic landforms*. Englewood Cliffs NJ: Prentice Hall.

Bloom, A. L. 1983: Sea-level and coastal morphology of the United States through the Late Wisconsin glacial maximum. In S. C. Porter (ed.), *Late Quaternary Environments of the United States*. London: Longman, 215–29.

Bolt, B. A. and Marion, W. C. 1966: Instrumental measurement of slippage on the Hayward Fault. *Seismological Society American Bulletin*, 56, 305–16.

Bonney, T. G. 1896: *Ice work present and past*. London: Kegan Paul, Trench and Trübner.

Bonney, T. G. 1902: Alpine valleys in relation to glaciers. *Quarterly Journal of the Geological Society*, 58, 690–702.

Bosscher, H. and Schlager, W. 1992: Computer simulation of reef growth. *Sedimentology*, 39, 503–12.

Boulaine, J. 1958: Sur la formation des carapaces calcaires. *Travaux des collaborateurs, Bulletin Service de la Carte Géologique de l'Algérie*, 20, 7–19.

Boulton, G. S. 1974: Processes and patterns of glacial erosion. In D. R. Coates (ed.), *Glacial Geomorphology*. London: Allen and Unwin, 41–87.

Boyé, M., Marmier, F., Nesson, C., and Trécolle, G. 1978: Les dépôts de la Sebkha Mellala. *Revue de Géomorphologie Dynamique*, 27, 49–62.

Bradley, W. C. and Griggs, G. B. 1976: Form, genesis and deformation of Central California wave-cut platform. *Geological Society America Bulletin*, 87, 433–49.

Branca, M. and Voltaggio. 1993: Erosion rate in badlands of central Italy: estimation by radiocaesium isotope ratios from Chernobyl nuclear accident. *Applied Geochemistry*, 8, 437–45.

Braun, D. D. 1989: Glacial and periglacial erosion of the Appalachians. *Geomorphology*, 2, 233–56.

Briat, M., Royer, A., Petit, J. R. and Lorius, C. 1982: Late glacial input of eolian continental dust in the Dome C Ice Core: additional evidence from individual microparticle evidence. *Annals of Glaciology*, 3, 27–31.

Bridge, J. S. and Leeder, M. R. 1979: A simulation model of alluvial stratigraphy. *Sedimentology*, 26, 617–44.

British Geomorphological Research Group, 1965: Unpublished symposium on rates of denudation.

Bromley, R. G. 1978: Comparative analysis of fossil and recent bioerosion. *Palaeontology*, 18, 725–39.

Brookfield, M. E. 1993: Miocene to Holocene uplift and sedimentation in the northwestern Himalaya and adjacent areas. In J. F. Shroder (ed.), *Himalaya to the sea*. London: Routledge, 43–71.

Brown, A. G. 1987: Holocene floodplain sedimentation and channel response of the lower River Severn, United Kingdom. *Zeitschrift für Geomorphologie*, 31, 293–310.

Brown, M. J., Krauss, R. K. and Smith, R. M. 1968. Dust deposition and weather. *Weatherwise*, 21(2), 66–9.

Bruijnzeel, L. A. 1983: The chemical mass balance of a small basin in a wet monsoonal environment and the effect of fast-growing plantation forest. *International Association for Scientific Hydrology, Publication*, 141, 229–39.

Brunsden, D. and Kesel, R. H. 1973: Slope development on a Mississippi River bluff in historic time. *Journal of Geology*, 81, 576–97.

Bucher, A. and Lucas, C. 1984: Sédimentation éolienne intercontinentale, poussières sahariennes et géologie. *Bulletin Centre Recherche Exploration Production Elf Aquitaine*, 8, 151–65.

Buddemeer, R. W. and Smith, S. V. 1988: Coral reef growth in an era of rapidly rising sea-level: predictions and suggestions for long-term research. *Coral Reefs*, 7, 51–6.

Bull, W. B. 1991: *Geomorphic responses to climatic change*. New York: Oxford University Press.

Bunte, K. and Poesen, J. 1994: Effect of rock fragment size and cover on overland flow hydraulics, local turbulence and sediment yield on an erodible soil surface. *Earth Surface Processes and Landforms*, 19, 115–35.

Burbank, D. W. and Beck, R. A. 1991: Rapid, long-term rates of denudation. *Geology*, 19, 1169–72.

Burnett, A. W. and Schumm, S. A. 1983: Alluvial-river response to neotectonic deformation in Louisiana and Mississippi. *Science*, 222, 49–50.

Burt, T. P., Donohoe, M. A. and Vann, A. R. 1983: The effect of forestry drainage operations on upland sediment yields: The results of a storm-based study. *Earth Surface Processes and Landforms*, 8, 339–46.

Butcher, D., Labadz, J. C., Botter, A. W. R. and White, P. 1993: Reservoir sedimentation rates in Southern Pennine region, UK. In J. McManus and R. W. Duck (eds), *Geomorphology and Sedimentology of Lakes and Reservoirs*. Chichester: Wiley, pp. 73–92.

Caine, N. 1976: A uniform measure of subaerial erosion. *Bulletin Geological Society of America*, 87, 137–40.

Caine, N. 1979: Rock weathering rates at the soil surface in an Alpine environment. *Catena*, 6, 131–44.

Caine, N. 1992: Spatial patterns of geochemical denudation in a Colorado alpine environment. In J. C. Dixon and A. D. Abrahams (eds), *Periglacial Geomorphology*, New York: Wiley, 63–88.

Cameron, C. C. 1970: Peat deposits of north-eastern Pennsylvania. *US Geological Survey Bulletin*, 1317A.

Carling, P. A. 1983: Thresholds of coarse sediment transport in broad and narrow natural streams. *Earth Surface Processes and Landforms*, 8, 1–18.

Carrara, P. E. and Carroll, T. R. 1979: The determination of erosion rates from exposed tree roots in the Piceance Basin, Colorado. *Earth Surface Processes*, 4, 307–17.

Carson, M. A. 1967: The magnitude of variability in samples of certain geomorphic characteristics drawn from valleyside slopes. *Journal of Geology*, 75, 93–100.

Carson, M. A. and Maclean, P. A. 1986: Development of hybrid aeolian dunes: the William River dune field, northwest Saskatchewan, Canada. *Canadian Journal of Earth Sciences*, 23, 1974–90.

Carter, L. D., Heginbottom, J. A. and Ming-ko Woo, 1987: Arctic Lowlands. *Geological Society of America Centennial Special Volume*, no. 2, 583–600.

Cernohouz, J. and Solc, I. 1966: Uses of sandstone rinds and weathered basaltic crust in absolute chronology. *Nature*, 212, 806–7.

Chadwick, W. W., Swanson, D. A., Iwatsubo, E. Y., Helicker, C. C. and Leighley, T. A. 1983: Deformation monitoring at Mount St. Helens in 1981 and 1982. *Science*, 221, 1378–80.

Chapman, V. J. and Ronaldson, J. W. 1958: The mangrove and saltmarsh flats of the Auckland Isthmus. *NZ Department of Scientific and Industrial Research, Bulletin*, 125, 79.

Chappell, J. and Polach, H. 1991: Post-glacial sea-level rise from a coral record at Huon Peninsula, Papua New Guinea. *Nature*, 349, 147–9.

Charles, G. 1949: Sur la formation de la carapace zonaire en Algérie. *Comptes rendus Académie des Sciences, Paris*, 228, 261–3.

Chinn, T. J. H. 1981: Use of rock weathering rind thickness for Holocene absolute age-dating in New Zealand. *Arctic and Alpine Research*, 13, 33–45.

Chorley, R. J., Dunn, A. J. and Beckinsale, R. P. 1964: *History of the Study of Landforms*, vol. 1. London: Methuen.

Chorley, R. J., Schumm, S. A. and Sugden, D. E. 1984: *Geomorphology*. London: Methuen.

Church, M. and Slaymaker, O. 1989: Disequilibrium of Holocene sediment yield in glaciated British Columbia. *Nature*, 337, 452–4.

Clark, M. and Small, J. 1982: *Slopes and weathering*. Cambridge: Cambridge University Press.

Clark, S. P. and Jäger, E. 1969: Denudation rate in the Alps from geochronologic and heat flow data. *American Journal of Science*, 267, 1143–60.

Clark, M. M., Grantz, A. and Rubin, M. 1972: Holocene activity of the Coyote Creek Fault as recorded in sediments of Lake Cahuilla. *US Geological Survey Professional Paper*, 787, 112–30.

Clarke, R. T. and McCulloch, J. S. G. 1979: The effect of land use on the hydrology of small upland catchments. In G. E. Hollis (ed.), *Man's impact on the hydrological cycle in The United Kingdom*. Norwich: Geo-Abstracts, 71–8.

Clemens, S. C. and Prell, W. L. 1990: Late Pleistocene variability of Arabian sea summer monsoon winds and continental aridity: eolian records from the lithogenic component of deep sea sediments. *Paleoceanography*, 5(2), 109–45.

Clemens, S. C., Farrell, J. W. and Gromet, L. P. 1993: Synchronous changes in seawater strontium isotope composition and global climate. *Nature*, 363, 607–10.

Clymo, R. S. 1991: Peat growth. In L. C. K. Shane and E. J. Cushing (eds), *Quaternary Landscapes*. London: Belhaven Press, 76–112.

Cole, K. L. and Mayer, L. 1982: Use of packrat middens to determine rates of cliff retreat in the eastern Grand Canyon, Arizona. *Geology*, 10, 597–9.

Coleman, J. M. 1969: Brahmaputra River: Channel processes and sedimentation. *Sedimentary Geology*, 3, 129–239.

Colman, S. M. 1981: Rock weathering rates as functions of time. *Quaternary Research*, 15, 250–64.

Condie, K. C. 1989: *Plate tectonics and crustal evolution* (3rd edn). Oxford: Pergamon Press.

Cooke, R. U. 1970: Stone pavements in deserts. *Annals Association of American Geographers* 60, 560–77.

Cooke, R. U. and Warren, A. 1973: *Geomorphology in deserts*. Batsford, London.

Cooke, R. U., Inkpen, R. J. and Wiggs, G. F. S. (in press) Changing rate of weathering in polluted atmospheres of the United Kingdom. *Earth Surface Processes and Landforms*.

Cooke, R. U., Warren, A. and Goudie, A. S. 1993: *Desert Geomorphology*, London: UCL Press.

Cooper, W. G. G. 1936: The bauxite deposits of the Gold Coast. *Bulletin Gold Coast Geological Survey*.

Corbel, J. 1959: Érosion en terrain calcaire. *Annales de Géographie*, 68, 97–120.

Corbel, J. 1964: L'érosion terrestre, étude quantitative (méthodes, techniques, résultats). *Annales de Géographie* 73, 385–412.

Costa, J. 1975: Effects of agriculture on erosion and sedimentation in the Piedmont Province, Maryland. *Bulletin of the Geological Society of America*, 86, 1281–86.

Costa, J. E. 1984: Physical geomorphology of debris flows. In J. E. Costa and P. J. Fleisher (eds), *Developments and applications of geomorphology*. Berlin: Springer-Verlag, 268–317.

Crittenden, M. D. 1967: New data on the isostatic deformation of Lake Bonneville. *US Geological Survey Professional Paper*, 454E, 1–31.

Croll, J. 1875: *Climate and time*, London: Daldy, Isbister and Co.

Cronin, T. M. 1981: Rates and possible causes of neotectonic vertical crustal movements of the emerged southeastern United States Atlantic coastal plain. *Bulletin Geological Society of America*, 92, 812–33.

Crowther, J. 1983: A comparison of the rock tablet and water hardness methods of determining chemical erosion rates on karst surfaces. *Zeitschrift für Geomorphologie*, 27, 55–64.

Crowther, J. 1987: Ecological observations in tropical karst terrain, west Malaysia. III Dynamics of vegetation-soil-bedrock system. *Journal of Biogeography*, 14, 157–64.

Dalloni, M. 1951: Sur la genèse et l'âge des 'terrains à croûte' Nords Africains. *Colloques internationaux, Centre National de la Recherche Scientifique*, 35, 278–85.

D'Almeida, G. A. 1986: A model for Saharan dust transport. *Journal of Climate and Applied Meteorology*, 25, 903–16.

Danin, A. 1983: Weathering of limestone in Jerusalem by cyanobacteria. *Zeitschrift für Geomorphologie*, 27, 413–412.

Darlington, J. P. E. C. 1985: Lenticular soil mounds in the Kenya highlands. *Oecologia*, 66, 116–21.

Davies, T. A., May, W. W., Southam, J. R. and Worsley, T. R. 1977: Estimates of Cenozoic oceanic sedimentation rates. *Science*, 197, 53–5.

Dawson, A. C., Matthews, J. A. and Shakesby, R. A. 1987: Rock platform erosion on periglacial shores: a modern analogue of Pleistocene rock platforms in Britain. In J. Boardman (ed.), *Periglacial processes and landforms in Britain and Ireland*. Cambridge: Cambridge University Press.

Day, M. J., Leigh, C., and Young, A. 1980: Weathering of rock discs in temperate and tropical soils. *Zeitschrift für Geomorphologie Supplementband*, 35, 11–15.

Dearing, J. A. and Foster, I. D. L. 1993: Lake sediments and geomorphological processes: some thoughts. In J. McManus and R. W. Duck (eds), *Geomorphology and Sedimentology of Lakes and Reservoirs*, Chichester: Wiley, pp. 5–14.

Deng, O., Fengnin, S., Shilong, Z., Mengluan, L., Tichin, W., Weiqi, Z., Burchtal, B. S., Molnar, P. and Peizhen, Z. 1984: Active faulting and tectonics in the Ningzia-Hui Autonomous Region, China. *Journal of Geophysical Research*, 89, 4427–45.

Denny, C. S. 1956: Surficial geology and geomorphology of Potter County. *US Geological Survey Professional Paper*, 288, 1–72.

Dietz, R. S. 1961: Continent and ocean basin evolution by spreading of the sea floor. *Nature*, 190, 854–7.

Doherty, J. T. and Lyons, J. B. 1980: Mesozoic erosion rates in northern New England. *Bulletin Geological Society of America*, 91, 16–20.

Dohrenwend, J. C. 1987: Basin and Range. In W. F. Graf (ed.), *Geomorphic systems of North America*, Boulder: Geological Society of America Centennial Special Volume 2, 303–31.

Dolan, R. and Kimball, S. 1985: Map of coastal erosion and accretion. In *National Atlas of the United States of America*, Reston, Va.: United States Geological Survey.

Donnelly, T. W. 1982: Worldwide continental denudation and climatic deterioration during the late Tertiary: evidence from deep-sea sediments. *Geology*, 10, 451–4.

Doornhof, D. 1992: Surface subsidence in the Netherlands: the Groningen Gas Field. *Geologie en Mijnbouw*, 71, 119–30.

Doornkamp, J. C. and Han Mukang. 1985: Morphotectonic research in China and its application to earthquake prediction. *Progress in Physical Geography*, 9, 353–81.

Dorn, R. and Oberlander, T. M. 1982. Rock varnish. *Progress in Physical Geography*, 6, 317–67.

Dorn, R. I. and Phillips, F. M. 1991: Surface exposure dating: review and critical evaluation. *Physical Geography*, 12, 303–33.

Douglas, I. 1969: The efficiency of humid tropical denudation systems. *Transactions Institute of British Geographers*, 46, 1–16.

Dragovich, D. 1986: Weathering rates of marble in urban environments, eastern Australia. *Zeitschrift für Geomorphologie*, 30, 203–14.

Drever, J. I. and Zobrist, J. 1992: Chemical weathering of silicate rocks as a function of elevation in the southern Swiss Alps. *Geochimica et Cosmochimica Acta*, 56, 3209–16.

Drewry, D. 1986: *Glacial geologic processes*. London: Arnold.

Dunne, T. 1978: Rates of chemical denudation of silicate rocks in tropical catchments. *Nature*, 274, 244–6.

Eardley, A. J. 1966: Rates of denudation in the High Plateaus of south-western Utah. *Bulletin Geological Society of America*, 77, 777–80.

Edge, R. J., Baker, T. F. and Jeffries, G. 1981: Borehole tilt measurements: a periodic crustal tilt in an aseismic area. *Tectonophysics*, 71, 97–109.

Edmond, J. M. 1992: Himalayan tectonics, weathering processes, and the strontium isotope record in marine limestones. *Science*, 258, 1594–7.

Einstein, H. A. 1950: The bedload function for sediment transportation in open channel flows. *US Department of Agriculture Technical Bulletin*, 1026.

Eldridge, D. J. and Greene, R. S. B. 1994: Assessment of sediment yield by splash erosion on a semi-arid soil with varying cryptogam cover. *Journal of Arid Environments* 26 (in press).

Ellison, W. D. 1945: Some effects of raindrops and surface flow on soil erosion and infiltration. *Transactions American Geophysical Union*, 26, 415–29.

Ellison, J. C. and Stoddart, D. R. 1990: Mangrove ecosystem collapse during predicted sea-level rise: Holocene analogues and implications. *Journal of Coastal Research*, 7, 151–65.

Elvidge, C. D. 1979: Distribution and formation of desert varnish in Arizona. MSc. thesis, Arizona State University.

Embabi, N. S. 1982: Barchans of the Kharga Depression. In F. El-Baz and T. A. Maxwell (eds), *Desert landforms of Southwestern Egypt: a basis for comparison with Mars*. Washington D.C.: NASA, 141–55.

Embabi, N. S. 1986/7: Dune movement in the Kharga and Dakhla oases depressions, the Western Desert, Egypt. *Bulletin de la Société de Géographie d'Egypte*, 59–60, 35–70.

Embleton, C. and King, C. A. M. 1968: *Glacial and periglacial geomorphology*. London: Arnold.

Emery, K. O. 1960: Weathering of the Great Pyramid. *Journal of Sedimentary Petrology*, 30, 140–3.

Emery, K. O. and Uchupi, E. 1972: Western North Atlantic Ocean. *Memoir Association of American Petroleum Geologists*, 17, 1–532.

Emmett, W. W. 1984: Measurement of bedload in rivers. In R. F. Hadley and D. E. Walling (eds), *Erosion and sediment yield: some methods of measurement and modelling*. Norwich: Geobooks, 91–109.

Endrody-Younga, S. 1982: Dispersion and translocation of dune specialist tenebrionids in the Namib Sea. *Cimbebasia*, 5, 257–71.

Engel, C. G. and Sharp, R. P. 1958: Chemical data on desert varnish. *Bulletin Geological Society of America*, 69, 487–578.

Eronen, M. 1983: Late Weichselian and Holocene shore displacement in Finland. In D. E. Smith and A. G. Dawson (eds), *Shorelines and Isostosy*, London: Academic Press, pp. 183–207.

Evans, G. 1979: Quaternary transgressions and regressions. *Journal of the Geological Society*, 126, 125–32.

Ewing, A. L. 1885: An attempt to determine the amount and rate of chemical erosion taking place in the limestone (Calciferous to Trenton) Valley of Center Country, Pa., and hence applicable to similar regions throughout the Appalachian regions. *American Journal of Science*, 29, 29–31.

Ewing, M. and Ewing, J. 1962: Rate of salt-dome growth. *Bulletin American Association of Petroleum Geologists*, 46, 708–9.

Eyles, N., Sasseville, D. R. Slatt, R. M., and Rogerson, R. J. 1982: Geochemical denudation rates and solute transport mechanisms in a maritime temperate glacier basin. *Canadian Journal of Earth Science*, 19, 1570–81.

Fairbanks, R. G. 1989: A 17,000 year glacio-eustatic sea-level record: influence of glacial melting rates on the Younger Dryas event and deep ocean circulation. *Nature*, 342, 637–42.

Fairbridge, R. W. 1981: The concept of neotectonics: an introduction. *Zeitschrift für Geomorphologie Supplementband*, 40, vii–xii.

Fairbridge, R. W. 1983: Isostasy and eustasy. In D. E. Smith and A. G. Dawson (eds), *Shorelines and Isostasy*, London: Academic Press, 3–25.

Feddema, J. J. and Meierding, T. C. 1987: Marble weathering and air pollution in Philadelphia. *Atmospheric Environment*, 21, 143–57.

Ferguson, R. 1981: Channel form and channel changes. In J. Lewin (ed.), *British Rivers*. London: Allen and Unwin, 90–125.

Ferguson, R. I. 1987: Accuracy and precision of methods for estimating river loads. *Earth Surface Processes and Landforms*, 12, 95–104.

Fillenham, L. F. 1963: Holme Fen Post. *Geographical Journal*, 129, 502–3.

Finkel, H. J. 1959: The barchans of southern Peru. *Journal of Geology*, 67, 614–47.

Finlayson, B. L. 1981: Field measurement of soil creep. *Earth Surface Processes and Landforms*, 6, 35–48.

Finlayson, B. L. and Osmaston, H. A. 1977: An instrument system for measuring soil movement. *BGRG Technical Bulletin*, 19.

Flach, K. W., Nettleton, W. D., Gile, L. H. and Cady, J. G. 1969: Pedocementation: induration by silica, carbonates and sesquioxides in the Quaternary. *Soil Science*, 107(6), 442–53.

Flemming, N. C. 1969. Archaeological evidence for eustatic change of sea level and earth movements in the western Mediterranean during the last 2000 years. *Geological Society American Special Paper*, 109, 1–125.

Fletcher, C. H. 1992: Sea-level trends and physical consequences: application to the US shore. *Earth-Science Reviews*, 33, 73–109.

Ford, D. C. 1993: Karst in cold environments. In H. M. French and O. Slaymaker (eds), *Canada's cold environments*. Montreal and Kingston: McGill-Queens University Press.

Ford, T. D., Gascoyne, M. and Deck, J. S. 1983: Speleothem dates and Pleistocene chronology in the Peak District of Derbyshire. *Cave Science* 10, 103–15.

Forsyth, D. W. and Uyeda, S. 1975: On the relative importance of the driving force of plate motions. *Geophysical Journal of the Royal Astronomical Society*, 43, 163–200.

Foster, T. 1976: *Bushfire*. Sydney: Reed.

Foster, I. D. L., Dearing, J. A., Simpson, A., Carter, A. D. and Appleby, P. G. 1985: Lake catchment based studies of erosion and denudation in the Merevale Catchment, Warwickshire, UK. *Earth Surface Processes and Landforms*, 10, 45–68.

Fournier, F. 1960: *Climat et érosion: la relation entre l'érosion du sol par l'eau et les précipitations atmosphériques*. Paris: Presses Universitaires France.

François, L. M. and Walker, J. C. G. 1992: Modelling the Phanerozoic carbon cycle and climate: constraints from the $^{87}Sr/^{86}Sr$ isotopic ratio of seawater. *American Journal of Science*, 292, 81–135.

Fraser. G. S. and de Calles, P. G. 1992: Geomorphic controls on sediment accumulation at margins of foreland basins. *Basin Research* 4, 233–52.

French, J. R. 1993: Numerical simulation of vertical marsh growth and adjustment to accelerated sea-level rise, north Norfolk, U.K. *Earth Surface Processes and Landforms*, 18, 63–81.

French, J. R. and Spencer, T. 1993: Dynamics of sedimentation in a tide-dominated backbarrier salt marsh, Norfolk, UK. *Marine Geology*, 110, 315–31.

Fryberger, S. G. 1979: Dune forms and wind regime. *US Geological Survey Professional Paper*, 1052, 137–69.

Fryberger, S. G. and Ahlbrandt, T. S. 1979: Mechanism for the formation of eolian sand seas. *Zeitschrift für Geomorphologie*, 23, 440–60.

Fryberger, S. G., Al-Sari, A. M., Clipshan, T. J., Rizvi, S. A. R. and Al-Hinai, K. G. 1984: Wind sedimentation in the Jafurah sand sea, Saudi Arabia. *Sedimentology*, 31, 413–31.

Galay, W. J. 1983: Causes of river bed degradation. *Water Resources Research*, 19, 1057–90.

Gale, S. J. 1992: Long-term landscape evolution in Australia. *Earth Surface Processes and Landforms*, 17, 323–43.

Gardner, J. S. 1983: Accretion rates on some debris slopes in the Mt. Rae area, Canadian Rocky Mountains. *Earth Surface Processes and Landforms*, 8, 347–55.

Gardner, J. S. and Jones, N. K. 1993: Sediment transport and yield at the Raikot Glacier, Nanga Parbat, Punjab Himalaya in J. F. Shroder (ed.), *Himalaya to the Sea*. London: Routledge, pp. 43–71.

Gardner, L. R. 1968: The Quaternary geology of the Mogei Valley, Clark Country, Nevada. Unpublished Ph.D. thesis, Penn State University.

Gardner, T. W., Jorgensen, D. W., Shurean, C. and Lemieux, C. R. 1987: Geomorphic and tectonic process rates: effects of measured time interval. *Geology*, 15, 259–61.

Garwood, E. J. 1910: Features of Alpine scenery due to glacial protection. *Geographical Journal*, 36, 310–339.

Gascoyne, M., Ford, D. C. and Schwarcz, H. P. 1983: Rates of cave and land form development in the Yorkshire Dales from speleothem age dates. *Earth Surface Processes and Landforms*, 8, 557–68.

Gavish, E. and Friedman, G. M. 1969: Progressive diagenesis in Quaternary to late Tertiary carbonate sediments: sequence and time scale. *Journal of Sedimentary Petrology*, 39, 980–1006.

Geikie, A. 1868: On denudation now in progress. *Geological Magazine*, 5, 249–54.

Geikie, A. 1880: Rock weathering as illustrated in Edinburgh church yards. *Proceedings Royal Society of Edinburgh*, 1879/80, 518–32.

Geikie, A. 1882: *Geological sketches at home and abroad*. London: Macmillan.

Geikie, A. 1893: *Text-book of geology*, London: Macmillan. 3rd. edition.

Gelinas, P. J. and Quigley R. M. 1973: The influence of geology on erosion rates along the north shore of Lake Erie. *Proceedings, 16th Conference on Great Lakes Research*, 421–30.

Gerson, R. and Amit, R. 1987: Rates and modes of dust accretion and deposition in an arid region – the Negev, Israel. In L. Frostick and I.

Reid (eds), *Desert sediments ancient and modern*. Oxford: Blackwell Scientific, 157–69.

Gibbs, R. J. 1967: The factors that control the salinity and composition of the suspended solids in the Amazon River. *Bulletin Geological Society of America*, 78, 1203–22.

Gilchrist, A. R. and Summerfield, M. A. 1991: Denudation, isostasy and landscape evolution. *Earth Surface Processes and Landforms*, 16, 555–62.

Gile, L. H. and Grossman, R. B. 1979: *The Desert Project Soil Monograph*. USA Soil Conservation Service.

Gill, E. D. 1981: Rapid honeycomb weathering (tafoni formation) in greywacke, S.E. Australia. *Earth Surface Processes and Landforms*, 6, 81–83.

Gill, E. D. and Lang, J. G. 1983: Micro-erosion metre measurements of rock wear on the Otway coast of southeast Australia. *Marine Geology*, 52, 141–56.

Gillette, D. 1980: Major contributions of natural primary continental aerosols: source mechanisms. *Annals New York Academy of Sciences*, 338, 348–58.

Gilluly, J. 1964: Atlantic sediments, erosion rates and the evolution of the continental shelf: some speculations. *Bulletin Geological Society of America*, 75, 483–92.

Glymph, L. M. 1951: Relation of sedimentation to accelerated erosion in the Missouri River Basin. *USDA Soil Conservation Service Technical Paper*, 103, 23.

Goldthwait, R. P. 1974: Rates of formation of glacier features in Glacier Bay, Alaska. In D. R. Coates (ed.), *Glacial Geomorphology*, London: Allen and Unwin, 163–85.

Goodchild, J. G. 1890: Notes on some observed rates of weathering of limestone. *Geological Magazine*, 27, 463–6.

Goudie, A. S. 1970: Input and output consideration in estimating rates of chemical denudation. *Earth Science Journal*, 4(2), 59–65.

Goudie, A. S. 1972: The concept of post-glacial progressive desiccation. *University of Oxford, School of Geography, Research Paper 4*.

Goudie, A. S. 1973: *Duricrusts in tropical and subtropical landscapes*, Oxford: Clarendon Press.

Goudie, A. S. 1978: Dust storms and their geomorphological implications. *Journal of Arid Environments*, 1, 291–310.

Goudie, A. S. 1983: Dust storms in space and time. *Progress in Physical Geography*, 7, 502–30.

Goudie, A. S. 1984: Salt Weathering. *School of Geography Research Paper Series*, 32.

Goudie, A. S. 1988: The geomorphological role of termites and earthworms in the tropics. In H. A. Viles (ed.), *Biogeomorphology*. Oxford: Blackwell, 166–92.

Goudie, A. S. 1989: Salt tectonics and geomorphology. *Progress in Physical Geography*, 13, 597–605.

Goudie, A. S. 1990: The global geomorphological future. *Zeitschrift für Geomorphologie Supplementband*, 79, 51–62.

Goudie, A. S. 1990: *The Landforms of England and Wales*. Oxford: Blackwell.

Goudie, A. S. (ed.) 1990: *Geomorphological techniques*. London: Unwin Hyman.

Goudie, A. S. 1993: *The human impact on the environment*. Oxford: Blackwell.

Goudie, A. S. and Thomas, D. S. G. 1985: Pans in southern Africa with particular reference to South Africa and Zimbabwe. *Zeitschrift für Geomorphologie*, 29, 1–19.

Goudie, A. S., Viles, H., Allison, R., Day, M., Livingstone, I. and Bull, P. A. 1990: The geomorphology of the Napier Range, Western Australia. *Transactions, Institute of British Geographers*, 15, 308–22.

Goudie, A. S., Stokes, S., Livingstone, I., Baliff, I. and Allison, R. J. 1993: Post-depositional modification of the Linear Sand ridges of the west Kimberley area of north-west Australia. *The Geographical Journal*, 159, 306–17.

Goudie, A. S., Viles, H. A. and Pentecost, A. 1993: The late-Holocene tufa decline in Europe. *The Holocene*, 3, 181–186.

Grant, D. R. 1980: Quaternary sea level change in Atlantic Canada as an indication of coastal delevelling. In N-A. Mörner (ed.), *Earth rheology, isostasy and eustasy*, 201–14, Chichester: Wiley.

Greenland, D. J. and Lal, R. 1977: *Soil conservation and management in the humid tropics*. Chichester: Wiley.

Gregory, J. W. 1913: *The nature and origin of fiords*. London: John Murray.

Griffiths, G. A. 1979: High sediment yields from major rivers of the western Southern Alps, New Zealand. *Nature*, 287, 88–9.

Griggs, D. T. 1936: The factor of fatigue in rock exfoliation. *Journal of Geology*, 44, 783–96.

Grigoryev, A. A. and Kondratyev, K. J. 1980: Atmospheric dust observed from space. *WMO Bulletin*, 3–9.

Grisez, L. 1960: Alvéolisation littorale de schistes métamorphiques. *Revue de Géomorphologie Dynamique*, 11, 164–7.

Gustavson, T. C. and Finley, R. J. 1985: Late Cenozoic geomorphic evolution of the Texas Panhandle and northeastern New Mexico. *Bureau of Economic Geology, University of Texas at Austin, Report of Investigations*, no. 148, 42 pp.

Hafsten, V. 1983: Biostratigraphical evidence for late Weichselian and Holocene sea-level changes in southern Norway. In D. E. Smith and A. G. Dawson (eds), *Shorelines and Isostasy*. London: Academic Press, pp. 161–81.

Haigh, M. J. 1977: The use of erosion pins in the study of slope evolution. *British Geomorphological Research Group Technical Bulletin*, 18, 31–49.

Hales, A. L. 1992: Speculations about crustal evolution. *Journal of Geodynamics*, 16, 55–64.

Hall, K. and Otte, W. 1990: A note on biological weathering of nunataks of the Juneau icefield, Alaska. *Permafrost and Periglacial Processes*, 1, 189–96.

Harbor, J. and Warburton, J. 1992: Glaciation and denudation rates. *Nature*, 356, 751.

Harrison, E. Z. and Bloom, A. L. 1977: Sedimentation rates on tidal salt marshes of Connecticut. *Journal of Sedimentary Petrology*, 47, 1484–90.

Harrison, J. B. 1911: On the formation of a laterite from a practically quartz-free diabase. *Geological Magazine*, 8, 120–3.

Hastenrath. S. L. 1967: The barchans of the Arequipa Region, southern Peru. *Zeitschrift für Geomorphologie*, NF 11, 300–331.

Hazelhoff, L., van Hoof, P., Imerson, A. C. and Kwaad, F. J. P. M. 1981: The exposure of forest soil to erosion by earthworms. *Earth Surface Processes and Landforms*, 6, 235–50.

Heinemann, H. G. 1984: Reservoir trap efficiency. In R. F. Hadley and D. E. Walling (eds), *Erosion and sediment yield: some methods of measurement and modelling*. Norwich: Geobooks, 201–18.

Hewitt, K. 1972: The mountain environment and geomorphic processes. In O. Slaymaker and H. J. McPherson (eds), *Mountain Geomorphology*. Vancouver: Tantalus Research, 17–34.

Hickox, C. F. 1959: Formation of ventifacts in a moist, temperate climate. *Geological Society of America Bulletin*, 70, 1489–90.

Hicks, D. M., McSaveney, M. J. and Chinn, T. J. H. 1990: Sedimentation in proglacial Ivory Lake, Southern Alps, New Zealand. *Arctic and Alpine Research*, 22, 26–42.

High, C. and Hanna, F. K. 1970: A method for the direct measurement of erosion on rock surfaces. *BGRG Technical Bulletin*, 5.

Hirschwald, J. 1908: *Die Prüfung der Natürlichen Bausteine auf ihre wetterbeständigkeit*. Berlin.

Hodell, D. A., Mead, G. A. and Mueller, P. A. 1990: Variation in strontium isotopic composition of seawater (8 Ma to present): Implications for chemical weathering rates and dissolved fluxes to the oceans. *Chemical Geology (Isotope Geosciences Section)*, 80, 291–307.

Hodgkin, E. P. 1964: Rate of erosion of intertidal limestone. *Zeitschrift für Geomorphologie*, 8, 385–92.

Holeman, J. N. 1968: The sediment yield of major rivers of the world. *Water Resources Research* 4, 737–47.

Holmes, A. 1965: *Principles of Physical Geology*. Edinburgh: Nelson.

Holt, J. A., Coventry, R. J. and Sinclair, D. F. 1980: Some aspects of the biology and pedological significance of mound-building termites in a red and yellow earth landscape near Charters Towers, north Queensland. *Australian Journal of Soil Research*, 18, 97–109.

Hooke, J. M. 1980: Magnitude and distribution of rates of river bank erosion. *Earth Surface Processes*, 5, 143–57.

Hooke, J. M. and Kain, R. J. P. 1982: *Historical change in the physical environment: a guide to sources and techniques*. London: Butterworth.

Horam, K. 1982: A satellite altimetric geoid in the Philippine Sea. *Nature*, 299, 117–21.

Houghton, J. T., Jenkins, G. J. and Ephraums, J. J. (eds) 1990: *Climate change: the IPCC scientific assessment*. Cambridge: Cambridge University Press.

Howard, A. D. 1992: Modelling channel migration and floodplain sedimentation in meandering streams. In P. A. Carling and G. E. Petts (eds), *Lowland Floodplain Rivers: Geomorphological Perspectives*. Chichester: Wiley. 1–41.

Howard, A. D. 1994: Slopes in layered rocks. In A. D. Abrahams and A. J. Parsons (eds), *Geomorphology of Desert Environments*, London: Chapman and Hall, 135–72.

Howell, D. G. 1989: *Tectonics of suspect terrains*. London: Chapman and Hall.

Hughes, R. J., Sullivan, M. E. and Yok, D. 1991: Human-induced erosion in a Highlands catchment in Papua New Guinea: the prehistoric and contemporary records. *Zeitschrift für Geomorphologie Supplementband*, 83, 227–239.

Hunt, C. B. and Mabey, D. R. 1966: Stratigraphy and structure, Death Valley, California. *US Geological Survey Professional Paper*, 494 A.

Hutchinson, J. N. 1970: A coastal mudflow on the London Clay Cliffs at Beltinge, north Kent. *Geotechnique*, 20, 412–38.

Imeson, A. C. 1971: Heather burning and soil erosion on the North Yorkshire Moors. *Journal of Applied Ecology*, 8, 537–41.

Inbar, M. 1992: Rates of fluvial erosion in basins with a Mediterranean climate type. *Catena*, 19, 393–409.

Inman, D. L., Ewing, G. C. and Corless, J. B. 1966: Coastal sand dunes of Guerrero Negro, Baja California, Mexico. *Bulletin Geological Society of America*, 77, 787–802.

Inoue, K. and Naruse, T. 1987: Physical, chemical and mineralogical characteristics of modern aeolian dust deposition. *Soil Science and Plant Nutrition*, 33, 327–45.

Jahn, A. 1961: Quantitative analysis of some periglacial processes in Spitzbergen. *Zesz. Nauk. Univ. Wroclaw*, ser. B., 5, 1–54.

Jahn, A. 1989: The soil creep on slopes in different altitudinal and ecological zones of Sudetes Mountains. *Geografiska Annaler*, 71A, 161–70.

Janda, R. J. 1971: An evaluation of procedures used in computing chemical denudation rates. *Bulletin Geological Society of America*, 82, 67–80.

Jansson, M. B. 1982: *Land erosion by water in different climates*. Uppsala University Naturgeog. Inst. Rapport, 57.

Jaynes, S. M. and Cooke, R. U. 1987: Stone weathering in south-east England. *Atmospheric Environment*, 21, 1601–22.

Jennings, J. N. 1985: *Karst Geomorphology*. Oxford: Blackwell.

Jiongxin Xu, 1993: A study of long-term environmental effects of river regulation on the Yellow River of China in Historical Perspective. *Geografiska Annaler*, 75, 61–72.

Johnston, M. J. S., McHugh, S. and Burford, R. D. 1976: On simultaneous tilt and creep observations on the San Andreas Fault. *Nature*, 260, 691–3.

Jones, R., Benson-Evans, K. and Chambers, F. M. 1985: Human influence upon sedimentation in Llangorse Lake, Wales. *Earth Surface Processes and Landforms*, 10, 227–35.

Jorgensen, D. W., Harvey, M. D., Schumm, S. A. and Flam, L. 1993: Morphology and dynamics of the Indus River: implications for the Mohenjo Daro site. In J. F. Shroder (ed.), *Himalaya to the Sea*. London: Routledge, 288–326.

Josens, G. 1983: The soil fauna of tropical savanna. III: The termites. In F. Bourlière (ed.), *Tropical savannas*. Amsterdam: Elsevier Scientific.

Joseph, J. H., Manes, A. and Ashbel, D. 1973: Desert aerosols transported by Khamsinic depressions and their climatic effects. *Journal of Applied Meteorology*, 12, 792–97.

Judson, S. 1968: Erosion rates near Rome, Italy. *Science*, 160, 1444–5.

Judson, S. and Ritter, D. F. 1964: Rates of regional denudation in the United States. *Journal of Geophysical Research*, 69, 3395–401.

Kaiser, E. 1926: *Die Diamantenwüste Südwestafrikas*, vol. 1. Berlin: Reimer.

Kaye, C. A. 1959: Shoreline features and Quaternary shoreline changes in Puerto Rico. *US Geological Survey Professional Paper*, 317, 1–140.

Kearey, P. and Vine, F. J. 1990: *Global tectonics*. Oxford: Blackwell Scientific.

Kelletat, D. 1991: The 1550 BP tectonic event in the Eastern Mediterranean as a basis for assessing the intensity of shore processes. *Zeitschrift für Geomorphologie Supplementband*, 81, 181–94.

Kenyon, P. M. and Turcotte, D. L. 1985: Morphology of a delta prograding by bulk sediment transport. *Bulletin Geological Society of America*, 96, 1457–65.

Keyes, 1913: Great erosional work of winds. *Popular Science Monthly*, May, 468–77.

Kiene, W. E. 1985: Biological destruction of experimental coral substrates at Lizard Island, Great Barrier Reef, Australia. *Proceedings Fifth International Coral Reef Congress*, 5, 339–44.

Kirk, R. M. 1977: Rates and forms of erosion on intertidal platforms at Kaikoura peninsula, South Island, New Zealand. *NZ Journal of Geology and Geophysics*, 20, 571–613.

Klein, M. 1984: Weathering rates of limestone tombstones measured in Haifa, Israel. *Zeitschrift für Geomorphologie*, 28, 105–11.

Kleman, J. 1994: Preservation of landforms under ice sheets and ice caps. *Geomorphology*, 9, 19–32.

Knox, J. C. 1989: Long and short-term episodic storage and removal of sediment in watersheds of southwestern Wisconsin and northwestern Illinois. *IAHS Publication*, 184, 157–64.

Knuepfer, P. L. K. 1992: Temporal variations in latest Quaternary slip across the Australian-Pacific Plate Boundary, Northeastern South Island, New Zealand. *Tectonics*, 11, 449–64.

Kolla, V. and Biscaye, P. E. 1977: Distribution and origin of quartz in the sediments of the Indian Ocean. *Journal of Sedimentary Petrology*, 47, 642–9.

Kostrzewski, A. and Zwolinski, Z. 1985: Chemical denudation rate in the upper Parseta catchment, western Pomerania: research methods and preliminary results. *Quaest Geog* (Pozan) 1, Special Issue, 121–38.

Kroger, P. M., Davidson, J. M. and Gardner, E. C. 1986: Mobile very long baseline interferometry and global positioning system measurement of vertical crustal motion. *Journal of Geophysical Research*, 91(B9), 9169–76.

Kuenen, Ph.H. 1956: Experimental abrasion of pebbles, 2. Rolling by current. *Journal of Geology*, 64, 336–68.

Kukal, Z. 1990: The rate of geological processes. *Earth-Science Reviews*, 28, 1–284.

Labadz, J. C., Burt, T. P. and Potter, A. W. R. 1991: Sediment yield and delivery in the blanket peat moorlands of the southern Pennines. *Earth Surface Processes and Landforms*, 16, 225–71.

Lal, D. 1991: Cosmic ray labelling of erosion surfaces: *in situ* nuclide production rates and erosion models. *Earth and Planetary Science Letters*, 104, 424–39.

Lam, K. C. 1977: Patterns and rates of slopewash on the badlands of Hong Kong. *Earth Surface Processes*, 2, 319–32.

Lamb, H. H. and Woodroffe, A. 1970: Atmospheric circulation during the last Ice Age. *Quaternary Research*, 1, 29–58.

Lambeck, K. 1988: *Geological geodesy*. Oxford: Clarendon Press.

Lambert, C. P. and Walling, D. E. 1987: Floodplain sedimentation: a preliminary investigation of contemporary deposition within the lower reaches of the River Culm, Devon, UK. *Geografiska Annaler*, 69A, 393–404.

Lancaster, N. 1989: *The Namib Sand Sea*. Rotterdam: Balkema.

Langbein, W. B. and Schumm, S. A. 1958: Yield of sediment in relation to mean annual precipitation. *Transactions American Geophysical Union*, 39, 1076–84.

Laronne, J. B. and Reid, I. 1993: Very high rates of bedload sediment transport by ephemeral desert rivers. *Nature*, 336, 148–50.

Larsen, E. and Mangerud, J. 1981: Erosion rate of a younger Dryas cirque glacier at Kråkenes, Western Norway. *Annals of Glaciology*, 2, 153–8.

Lasaga 1984: Chemical kinetics of water rock interactions. *Journal of Geophysical Research*, 89(B6), 4009–25.

Lawler, D. 1986: River bank erosion and the influence of frost: a statistical examination. *Transactions, Institute of British Geographers*, 11, 227–42.

Lawler, D. M. 1991: A new technique for the automatic monitoring of erosion and deposition rates. *Water Resources Research*, 27, 2125–28.

Lee, K. E. and Wood, T. G. 1971a: *Termites and soils*. London and New York: Academic Press.

Lee, K. E. and Wood, T. G. 1971b: Physical and chemical effects on soils of some Australian termites and their pedological significance. *Pedobiologia*, 11, 376–409.

Leinen, M. 1989: The late Quaternary record of atmospheric transport to the northwest Pacific from Asia. In M. Leinen and M. Sarnthein (eds), *Palaeoclimatology and Palaeometeorology: Modern and Past Patterns of Global Atmospheric Transport*, Kluwer: Dordrecht, 693–732.

Leinen, M. and Heath, G. R. 1981: Sedimentary indicators of atmospheric activity in the northern hemisphere during the Cenozoic. *Palaeogeography, Palaeoclimatology, Palaeoecology*, 36, 1–21.

Leopold, L. B., Wolman, M. G. and Miller, J. P. 1964: *Fluvial processes in Geomorphology*. San Francisco: Freeman.

Lepage, M. 1974: Les termites d'une savane sahélienne (Ferlo Septentrional, Sénégal): peuplement, populations, consommation, rôle dans l'écosystème. D.Sc. Thesis. University of Dijon.

Lepage, M. 1984: Distribution, density and evolution of *Macrotermes bellicosus* nests (Isoptera: Macrotermitinae) in the north-east of the Ivory Coast. *Journal of Animal Ecology*, 53, 107–17.

Lewin, J., Bradley, S. B. and Macklin, M. G. 1983: Historical valley alluviation in mid-Wales. *Geological Journal*, 18, 331–50.

Lewis, C. L. E., Green, P. F., Carter, A. and Hurford, A. J. 1992: Elevated K/T palaeotemperatures throughout Northwest England: three kilometres of Tertiary erosion. *Earth and Planetary Science Letters*, 112, 131–45.

Lewkowicz, A. G. 1987: Nature and importance of thermokarst processes, Sand Hill Moraine, Banks Island, Canada. *Geografiska Annaler*, 69A, 321–7.

Lewkowicz, A. G. 1988: Slope processes. In M. J. Clark (ed.), *Advances in periglacial geomorphology*. Wiley: Chichester, 325–68.

Lisitzin, E. 1974: *Sea-level changes*. Amsterdam: Elsevier.

Livingstone, I. 1989: Monitoring surface change on a Namib linear dune. *Earth Surface Processes and Landforms*, 14, 317–32.

Long, J. T. and Sharp, R. P. 1964: Barchan-dune movement in Imperial Valley, California. *Bulletin Geological Society of America*, 75, 149–56.

Loÿe-Pilot, M. D., Martin, J. M. and Morelli, J. 1986: Influence of Saharan dust on the rain acidity and atmospheric input to the Mediterranean. *Nature*, 321, 427–8.

Luckman, B. H. 1978: Geomorphic work of snow avalanches in the Canadian Rocky Mountains. *Arctic and Alpine Research*, 10, 261–76.

McCool, D. K., Dossett, M. G. and Yecha, S. J. 1981: A portable rill meter for field measurement of soil loss. *Publication International Association of Hydrological Sciences*, 13, 479–84.

Mackay, J. R. 1978: Contemporary pingos: a discussion. *Biulteyn Pery-glacjalny*, 27, 133–54.

Mackintosh, D. 1869: *The Scenery of England and Wales, its character and origin*, London: Longman, Green.

Magee, A. W., Bull, P. A. and Goudie, A. S. 1988: Chemical textures on quartz grains: an experimental approach using salts. *Earth Surface Processes and Landforms*, 13, 665–76.

Magilligan, F. J. 1985: Historical floodplain sedimentation in the Galena River basin, Wisconsin and Illinois. *Annals of the Association of American Geographers*, 75, 583–94.

Maire, R. 1981: Les hauts karsts periméditerranéens. *Proceedings 8th International Congress of Speleology*, 788–92.

Maley, J. 1980: Les changements climatiques de la fin de Tertiaire en Afrique: leur conséquence sur l'apparition du Sahara et sa végétation. In M. A. J. Williams and H. Faure (eds), *The Sahara and the Nile*, Rotterdam: Balkema, 63–86.

Marsh, G. P. 1864: *Man and Nature*. New York: Scribner.

Matsukura, Y. and Matsuoka, N. 1991: Rates of tafoni weathering on uplifted shore platforms in Nomija Zaki, Boso Peninsula, Japan. *Earth Surface Processes and Landforms*, 16, 51–6.

Matthews, W. H. 1975: Cenozoic erosion and erosion surfaces of eastern North America. *American Journal of Science*, 275, 818–24.

Matthews, J. A. and Berrisford, M. S. 1993: Climatic controls on rates of solifluction: variations within Europe. In B. Frenzel, J. A. Matthews, and S. B. Gläser (eds), *Solifluction and climatic variation in the Holocene*, Stuttgart: Gustav Kischer, 363–382.

Matthias, G. F. 1967: Weathering rates of Portland arkose tombstones. *Journal of Geological Education*, 15, 140–4.

McCauley, J. F., Breed, C. G., El-Baz, F., Whitney, M. I., Grolier, M. J. and Ward, A. W. 1979: Pitted and fluted rocks in the western Desert of Egypt: Viking companions. *Journal of Geophysical Research*, 84, 8222–31.

McCauley, J. F., Grolier, M. J. and Breed, C. S. 1977: Yardangs of Peru and other desert regions. *US Geological Survey Interagency Report, Astrogeology*, 81, 177 pp.

McGee, W. J. 1897: Sheetflood erosion. *Bulletin Geological Society of America*, 8, 87–112.

McGee, W. J. 1911: Soil erosion. *US Department of Agriculture Bureau of Soils Bulletin* 71, 1–60.

McGuire, B. and Saunders, S. 1993: Recent earth movements at active volcanoes: a review. *Quaternary Proceedings*, 3, 33–46.

McLennan, S. M. 1993: Weathering and global denudation. *Journal of Geology*, 101, 295–303.

McManus, J. and Duck, R. W. 1985: Sediment yield estimated in the Ochil Hills, Scotland. *Earth Surface Processes and Landforms*, 10, 193–200.

McNutt, M. and Menard, H. W. 1978: Lithospheric flexure and uplifted atolls. *Journal of Geophysical Research*, 83(B3), 1206–12.

McTainsh, G. and Walker, P. H. 1982: Nature and distribution of Harmattan Dust. *Zeitschrift für Geomorphologie* NF 26, 417–36.

Meade, R. H. 1969: Errors in using modern stream-load data to estimate natural rates of denudation. *Bulletin Geological Society of America*, 80, 1265–74.

Meade, R. H. and Trimble, S. W. 1974: Changes in sediment loads in rivers of the Atlantic drainage of the United States since 1900. *IASH Publication*, 113, 99–104.

Meade, R. H. 1982: Sources, sinks and storage of river sediment in the Atlantic drainage of the United States. *Journal of Geology*, 90, 235–52.

Meade, R. H. and Parker, R. S. 1985: Sediment in rivers in the United States. *United States Geological Survey Water Supply Paper*, 2275, 49–60.

Meert, J. G., der Voo, R. V., Powell, C.McA., Zheng-Xiang, Li., McElhinny, M. W., Zhong Chen and Symons, D. T. A. 1993: A plate tectonic speed limit? *Nature*, 363, 216–17.

Meierding, T. C. 1993a: Inscription legibility method for estimating rock weathering rates. *Geomorphology*, 6, 273–86.

Meierding, T. C. 1993b: Marble tombstone weathering and air pollution in North America. *Annals of the Association American Geographers*, 83, 568–88.

Merrill, G. P. 1904: *Rocks, rock weathering and soils*, New York: Macmillan.

Meybeck, M. 1979: Concentrations des eaux fluviales en éléments majeurs et apports en solution aux océans. *Revue de Géologie Dynamique et de Géographie Physique*, 21, 215–46.

Meybeck, M. 1983: Atmospheric inputs and river transport of dissolved substances. *International Association of Hydrological Sciences Publication*, 141, 173–92.

Meybeck, M. 1987: Global chemical weathering of surficial rocks estimated from river dissolved loads. *American Journal of Science*, 287, 401–28.

Meyer, L. D. 1982: Soil erosion research leading to development of the Universal Soil Loss Equation. In *Proceedings Workshop on Estimating erosion and sediment yield on Rangelands*, 1–16. Tuscon, Arizona, 7–9 March 1881.

Miller, D. S. and Lakatos, S. 1983: Uplift rate of Adirondack anorthosite measured by fission-track analysis of apatite. *Geology*, 11, 284–6.

Mills, H. M. 1984: Effect of hillslope angle and substrate on tree tilt and denudation of hillslopes by tree fall. *Physical Geography*, 5, 253–61.

Milliman, J. D. 1990: Fluvial sediment in coastal seas: flux and fate. *Nature and Resources*, 26, 12–22.

Milliman, J. D. 1991: Flux and fate of fluvial sediment and water in coastal seas. In R. F. C. Mantouva, J-M. Martin and R. Wollest (eds), *Oceanic margin processes in coastal change*, Chichester: Wiley.

Milliman, J. D., Qin, Y. S., Ren, M. E. and Yoshiki Saita 1987: Man's influence on erosion and transport of sediment by Asian rivers: The Yellow River (Huanghe) example. *Journal of Geology*, 95, 751–62.

Milliman, J. D. and Syvitski, J. P. M. 1992: Geomorphic/tectonic control of sediment discharge to the Ocean: the importance of small mountainous rivers. *Journal of Geology*, 100, 525–44.

Mooney, H. A. and Parsons, D. J. 1973: Structure and function of the California Chaparral – an example from San Dimas. *Ecological Studies*, 7, 83–112.

Moore, T. R. 1979: Land use and erosion in the Machakos Hills. *Annals of the Association of American Geographers*, 69, 419–31.

Moore, J. G. and Clague, D. A. 1992: Volcano growth and evolution of the island of Hawaii. *Bulletin Geological Society of America*, 104, 1471–84.

Moore, P. D. and Bellamy, D. J. 1974: *Peatlands*. London: Elek Science.

Moore, R. J. and Newson, M. D. 1986: Production, storage and output of coarse upland sediments: natural and artificial influences as revealed by research catchment studies. *Journal of the Geological Society of London*, 143, 921–6.

Morgan, R. P. C. 1977: Soil erosion in the United Kingdom: field studies in the Silsoe area, 1973–75. *National College of Agricultural Engineering, Occasional Paper*, 4.

Morgan, R. P. C. 1986: *Soil erosion and conservation*. Harlow: Longman.

Mottershead, D. N. 1982: Coastal spray weathering of bedrock in the supratidal zone at East Prawle, South Devon. *Field Studies*, 5, 663–84.

Mottershead, D. N. 1992: Alveolar weathering by marine salt attack of sandstone in a dated structure: a study of Weston-Super-Mare sea walls. Paper presented at BGRG Annual Conference, University of Sussex, 27 September 1992.

Muhs, D. R. 1983: Airborne dustfall on the California Channel Islands, USA. *Journal of Arid Environments*, 6, 223–228.

Muhs, D. R., Rockwell, T. K. and Kennedy, G. L. 1992: Late Quaternary uplift rates of marine terraces on the Pacific coast of North America, southern Oregon to Baja California Sur. *Quaternary International*, 15–16, 121–33.

Muhs, D. R., Kelsey, H. M., Miller, G. H., Kennedy, G. L., Whelan, J. F. and McInelly, G. W. 1990: Age estimates and uplift rates for late Pleistocene marine terrace: southern Oregon portion of the Cascadia Forearc. *Journal of Geophysical Research*, 95(B) 6685–98.

Nahon, D. B. 1991: *Introduction to the petrology of soils and chemical weathering*. New York: Wiley.

Neil, D. 1989: Weathering rates of a subaerially exposed marble in eastern Australia. *Zeitschrift für Geomorphologie*, 33, 463–73.

Nesje, A., Dahl, S. O., Valen, V. and Øvstedal, J. 1992: Quaternary erosion in the Sognefiord drainage basin, western Norway. *Geomorphology*, 5, 511–20.

Newson, M. D. 1981: Mountain streams. In J. Lewin (ed.), *British Rivers*, London: Allen and Unwin, 59–89.

Newson, M. D. 1986: River basin engineering – fluvial geomorphology. *Journal of the Institution of Water Engineers and Scientists*, 40, 309–25.

Nikonov, A. A. 1977: Contemporary technogenic movements of the Earth's crust. *International Geology Review*, 19, 1245–58.

Nikonov, A. A. and Yu Shebalina, T. 1979: Lichenometry and earthquake age determination in central Asia. *Nature*, 280, 675–7.

Nishiizumi, K., Kohl C. P., Arnold, J. R., Dorn, R., Klein, J., Fink, D., Middleton, R., and Lal, D., 1993: Role of in situ cosmogenic nuclides ^{10}Be and ^{26}Al in the study of diverse geomorphic processes. *Earth Surface Processes and Landforms*, 18, 407–25.

Norris, R. M. 1966: Barchan dunes of Imperial Valley, California. *Journal of Geology*, 74, 292–306.

Nyberg, R. 1987: Slush avalanche erosion along stream courses in the northern Swedish mountains. In A. Godard and A. Rapp (eds), *Processus et mesure de l'érosion*, Meudon-Bellevue: CNRS, 179–86.

Nye, P. H. 1955: Some soil forming processes in the humid tropics. IV: The action of the soil fauna. *Journal of Soil Science*, 6, 73–83.

Oberlander, T. M. 1994: Rock varnish in deserts. In A. D. Abrahams and A. J. Parsons (eds), *Geomorphology of Desert Environments*. London: Chapman and Hall, 106–19.

Oenema, O. and De Laure, R. D. 1988: Accretion rates in salt marshes in the eastern Scheldt, south-west Netherlands. *Estuarine, Coastal and Shelf Science*, 26, 379–94.

Oerlemans, J. 1994: Quantifying global warming from the retreat of glaciers, *Science*, 264, 243–5.

Ohmura, A. and Reeh, N. 1991: New precipitation and accumulation
maps for Greenland. *Journal of Glaciology*, 37, 140–8.

Oldfield. F. 1981: Peats and lake sediments: formation, stratigraphy,
description and nomenclature. In A. S. Goudie (ed.), *Geomorphological
Techniques*, London: Allen and Unwin, 306–26.

Olson, J. C. 1958: Rates of succession and soil changes on southern Lake
Michigan sand dunes. *Botanical Gazette*, 119, 125–70.

Orme, A. R. 1991: Mass movement and seacliff retreat along the southern California coast. *Bulletin Southern California Academy of Sciences*,
90, 58–79.

Ovenden, L. 1990: Peat accumulation in northern wetlands. *Quaternary
Research*, 33, 377–86.

Owens, P. and Slaymaker, O. 1992: Late Holocene sediment yields in
small alpine and subalpine drainage basins, British Columbia. *IAHS
Publication*, 209, 147–54.

Oxley, N. C. 1974: Suspended sediment delivery rates and solute concentration of stream discharge in two Welsh catchments. *Institute of
British Geographers' Special Publication*, 6, 141–53.

Parkin, D. W. and Shackleton, N. J. 1973: Trade winds and temperature
correlations down a deep-sea core off the Sahara coast, *Nature*, 245,
455–7.

Passarge, S. 1904: Die inselberglandschafte in tropischen Afrika. *Naturwiss,
Wochens*, 3, 657–65.

Paull, C. K., Neumann, A. C., Bebout, B., Zabielski and V. Showers, W.
1992: Growth rate and stable isotopic character of modern stromatolites
from San Salvador, Bahamas. *Palaeogeography, Palaeoclimatology,
Palaeoecology*, 95, 335–44.

Pavoni, N. and Green, R. (eds) 1975: *Recent crustal movements*. Developments in Geotectonics No. 9.

Penck, W. 1953: *Morphological analysis of land forms. A contribution to
physical geology*. London: Macmillan.

Pereira, H. C. 1973: *Land use and water resources in temperate and
tropical climates*. Cambridge: Cambridge University Press.

Persons, B. S. 1970: *Laterite, genesis, location, use*. New York.

Petersen, J. T. and Junge, C. E. 1971: Sources of particulate matter in
the atmosphere. In W. H. Matthews, W. W. Kellogg and G. D. Robinson,
Man's impact on the climate, Cambridge, Mass.; MIT Press, 310–20.

Pethick, J. S. 1981: Long-term accretion rates on tidal salt marshes. *Journal
of Sedimentary Petrology*, 51, 511–17.

Petts, G. E. 1979: Complex response of river channel morphology subsequent to reservoir construction. *Progress in Physical Geography*, 3,
329–62.

Péwé, T. L., Péwé, E. A. Péwé, R. H., Journaux, A. and Slatt, R. M. 1981:
Desert dust: characteristics and rates of deposition in central Arizona,
USA. *Geological Society of America Special Paper*, 186, 169–90.

Phillips, J. D. 1990: Relative importance of factors influencing fluvial soil
loss at the global scale. *American Journal of Science*, 290, 547–68.

Phillips, J. D. 1991: Fluvial sediment budgets in the North Carolina Piedmont. *Geomorphology*, 4, 231–42.

Pinet, P. and Souriau, M. 1988: Continental erosion and large-scale relief.
Tectonics 7, 563–82.

Pirazzoli, P. A. 1989: Past and near-future global sea-level changes. *Palaeogeography, Palaeoclimatology and Palaeoecology*, 75, 241–58.

Pirazzoli, P. A. 1991: *World Atlas of Holocene sea-level changes*. Amsterdam: Elsevier.

Pizzuto, J. E. 1987: Sediment diffusion during overbank flows. *Sedimentology*, 34, 301–17.

Playford, P. E. and Cockbain, A. E. 1976: Modern algal stromatolites at Hamelin Pool, a hypersaline barred basin in Shark Bay, Western Australia. *Developments in Sedimentology*, 20, 389–411.

Poesen, J. and Lavee, H. 1991: Effects of size and incorporation of synthetic mulch on runoff and sediment yield from interrills in a laboratory study with simulated rainfall. *Soil and Tillage Research*, 21, 209–23.

Pokras, E. M. 1989: Pliocene history of south Saharan/Sahelian aridity: record of freshwater diatoms (Genus *Melosira*) and opal phytoliths, ODP Sites 662 and 664. In M. Leinen and M. Sarnthein (eds), *Palaeoclimatology and Palaeometeorology: Modern and Past Patterns of Global Atmospheric Transport*, Kluwer: Dordrecht, 795–804.

Pollard, W. H. 1988: Seasonal frost mounds. In M. J. Clarke (ed.), *Advances in Periglacial Geomorphology*, Chichester: Wiley, 201–29.

Pomeroy, D. E. 1976: Some effects of mound-building termites on soils in Uganda. *Journal of Soil Science*, 27, 377–94.

Pouquet, J. 1966: *Initiation géopédologique*. Paris.

Press, F. and Siever, R. 1986: *Earth*. New York: Freeman.

Prospero, J. M. 1981: Arid regions as sources of mineral aerosols in the marine atmosphere. *Geological Society of America Special Paper*, 186, 71–86.

Pullan, R. A. 1979: Termite hills in Africa: their characteristics and evolution. *Catena* 6, 267–91.

Pye, K. 1987: *Aeolian dust and dust deposits*, London: Academic Press.

Rahns, P. H. 1971: The weathering of tombstones and its relationship to the topography of New England. *Journal of Geological Education*, 19, 112–18.

Ranwell, D. S. 1964: *Spartina* salt marshes in southern England: II. Rate and seasonal pattern of sediment accretion. *Journal of Geology*, 52, 79–94.

Rapp, A. 1960: Recent development of mountain slopes in Karkevagge and surroundings, northern Scandinavia. *Geografiska Annaler*, A, 42, 65–200.

Rapp, A. Murray-Rust, D. H., Christiansson, C. and Berry, L. 1972: Soil erosion and sedimentation in four catchments near Dodoma, Tanzania. Geografiska Annaler, 54A, 255–318.

Rausch, D. L. and Heinmann, H. G. 1984: Measurement of reservoir sedimentation. In R. F. Hadley and D. E. Walling (eds), *Erosion and sediment yield: some methods of measurement and modelling*. Norwich: Geobooks, 179–200.

Rea, D. K. 1989: Geologic record of atmospheric circulation on tectonic time scales. In M. Leinen and M. Sarnthein (eds), *Palaeoclimatology and Palaeometeorology: Modern and Past Patterns of Global Atmospheric Transport*, Kluwer: Dordrecht, 841–57.

Reade, T. M. 1885: Denudation of the two Americas. *American Journal of Science*, 29, 290–300.

Reches, Z. and Hoexter, O. F. 1981: Holocene seismic and tectonic activity in the Dead Sea area. *Tectonophysics* 80, 235–54.

Reclus, E. 1873: *The Ocean, atmosphere and life*. New York: Harper.

Reclus, E. 1881: *The history of a mountain*, London: Sampson Low, Marston, Searle and Rivington.

Reddy, M. M. 1988: Acid rain damage to carbonate stone: a quantitative assessment based on the aqueous geochemistry of rainfall runoff from stone. *Earth Surface Processes and Landforms*, 13, 335–54.

Reeckman, S. A. and Gill, E. D. 1981: Rates of vadose diagenesis in Quaternary dune and shallow marine carbonates, Warrnambool, Victoria, Australia. *Sedimentary Geology*, 30, 157–72.

Reed, D. J. 1988: Sediment dynamics and deposition in a retreating coastal marsh. *Estuarine, Coastal and Shelf Science*, 26, 67–69.

Reed, D. J. 1990: The impact of sea-level rise on coastal salt marshes. *Progress in Physical Geography*, 14, 465–81.

Reheis, M. J. 1975: Source, transporation and deposition of debris on Arapaho Glacier, Front Range, Colorado, USA. *Journal of Glaciology*, 14, 407–20.

Reneau, S. and Dietrich, W. E. 1991: Erosion rates in the southern Oregon Coast Range: evidence for an equilibrium between hillslope erosion and sediment yield. *Earth Surface Processes and Landforms*, 16, 307–22.

Reniger, A. 1955: Erozja gleb na terenie podogorskim w obrebie zlewin potokn Lukawica. *Roczu. Nauk. Roln, Ser. F.*, 71, 149–210.

Revelle, R. and Emery, K. O. 1957: Chemical erosion of beach rock and exposed reef rock. *United States Geological Survey Professional Paper*, 260T, 699–709.

Reynolds, B, 1986: A comparison of element outputs in solution, suspended sediments and bedload for a small upland catchment. *Earth Surface Processes and Landforms*, 11, 217–21.

Reynolds, R. C. and Johnson, N. M. 1972: Chemical weathering in the temperate glacial environment of the Northern Cascade Mountains. *Geochimica et Cosmochimica Acta*, 36, 537–54.

Richardson, S. J. and Smith, J. 1977: Peat wastage in the East Anglian Fens. *Journal of Soil Science*, 28, 485–9.

Riser, J. 1985: Le rôle du vent au cours des derniers millénaires dans le Bassin Saharien D'Araouane (Mali). *Bulletin de l'Association de Géographes Français*, 62, 311–17.

Ritchie, J. C., Spraberry, J. A. and McHenry, J. R. 1974: Estimating soil erosion from the redistribution of fallout 137 Cs. *Proceedings Soil Science Society of America*, 38, 137–9.

Rivera, A., Ledoux, E. and Marsily, G. de. 1991: Nonlinear modelling of groundwater flow and total subsidence of the Mexico City Aquifer-Aquitard System. *IAHS Publication 200*, 45–58.

Robinson, A. H. W. 1980: Erosion and accretion along part of the Suffolk coast of East Anglia, England. *Marine Geology*, 37, 133–46.

Robinson, L. A. 1977: Marine erosive processes at the cliff foot. *Marine Geology*, 23, 257–71.

Roels, J. M. 1985: Estimation of soil loss at a regional scale based on plot measurements – some critical considerations. *Earth Surface Processes and Landforms*, 10, 587–95.

Rowe, P. 1988: Rates of incision in central England during the Quaternary. *Quaternary Newsletter*, 56, 21–32.

Royal Commission on Coastal Erosion. 1911: Report. London: HMSO.

Rubin, D. M. 1990: Lateral migration of linear dunes in the Strezclecki Desert, Australia. *Earth Surface Processes and Landforms*, 15, 1–14.

Ruskin, J. 1865: Notes on the shape and structure of some parts of the Alps, with reference to denudation. *Geological Magazine*, 2, 49–54 and 193–6.

Ruxton, B. P. and McDougall, I. 1967: Denudation rates in northeast Papua from potassium-argon dating of lavas. *American Journal of Science*, 265, 545–61.

Sadler, P. M. 1981: Sediment accumulation rates and the completeness of the stratigraphic sections. *Journal of Geology*, 89, 569–84.

Safar, M. I. 1985: *Dust and dust storms in Kuwait*. State of Kuwait: Kuwait.

Salisbury, E. J. 1925: Note on the edaphic succession in some dune soils with special reference to the time factor. *Journal of Ecology*, 13, 322–8.

Sarnthein, M. and Walger, K. 1974: Der äolische sandstrom aus der w-sahara zür Atlantikkuste. *Geologische Rundschau*, 63, 1065–87.

Sarre, R. 1989: Aeolian sand drift from the intertidal zone on a temperate beach: potential and actual rates. *Earth Surface Processes and Landforms*, 14, 247–58.

Saunders, I. and Young, A. 1983: Rates of surface processes on slopes, slope retreat and denudation. *Earth Surface Processes and Landforms*, 8, 473–501.

Schachak, M., Jones, C. G. and Grand, Y. 1987: Herbivory in rocks and the weathering of a desert. *Science*, 236, 1098–99.

Schmidt, K-H. 1985: Regional variation of mechanical and chemical denudation, Upper Colorado River Basin, USA. *Earth Surface Processes and Landforms*, 10, 497–508.

Schmidt, K-H. 1988: Rates of scarp retreat: a means of dating Neotectonic activity. In V. H. Jacobshagen (ed.), *The Atlas System of Morocco*, Berlin: Springer Verlag, 445–62.

Schmidt, K-H. 1989: The significance of scarp retreat for Cenozoic Landform evolution on the Colorado Plateau, USA. *Earth Surface Processes and Landforms*, 14, 93–105.

Schramm, C. T. 1989: Cenozoic climatic variation recorded by quartz and clay minerals in North Pacific sediments. In M. Leinen and M. Sarnthein (eds), *Palaeoclimatology and Palaeometeorology: Modern and Past Patterns of Global Atmospheric Transport*, Kluwer: Dordrecht, 805–39.

Schumm, S. A. 1963: The disparity between present rates of denudation and orogeny. *US Geological Survey Professional Paper*, 454-H, 1–13.

Schumm, S. A. 1973: Abrasion in place: a mechanism for rounding and size reduction of coarse sediments in rivers. *Geology*, 1, 37–40.

Schumm, S. A. 1977: *The fluvial system*. New York: Wiley.

Schutz, L. 1980: Long-range transport of desert dust with special emphasis on the Sahara. *Annals New York Academy of Science*, 338, 515–32.

Schwartz, H. E., Enel, J., Dickens, W. J., Rogers, P. and Thompson, J. 1991: Water quality and flows. In B. L. Turner (ed.), *The Earth as transformed by Human Action*, Cambridge: Cambridge University Press, 253–70.

Selby, M. J. 1985: *Earth's changing surface*. Oxford: Oxford Unversity Press.

Seni, S. J. and Jackson, M. P. A. 1983: Evolution of salt structures, East Texas diapir province. Part 2, Patterns and rates of halokinesis. *Bulletin American Association of Petroleum Geologists*, 67, 1245–74.

Shehata, W., Bader, T., Irtem, O., Ali, A., Abdullah, M. and Aftab, S. 1992: Rate and mode of barchan dune advance in the central part of the Jafurah Sand Sea. *Journal of Arid Environments*, 23, 1–17.

Shaler, N. S. 1898: *Outline of the Earth's History*. London: Heinemann.

Shaler, N. J. 1912: *Man and the earth*. New York: Sheffield.

Shapiro, I. I. 1983: Use of space techniques for geodesy. In *Earthquakes: observations, theory and interpretation*, H. Kanamori and E. Boschi (eds), Amsterdam: North-Holland, 530–68.

Sharp, R. P. 1949: Pleistocene ventifacts east of the Big Horn Mountains, Wyoming. *Journal of Geology*, 57, 175–95.

Sharp, R. P. 1964: Wind-driven sand in Coachella Valley, California. *Bulletin Geological Society of America*, 75, 785–804.

Shennan, I. 1992: Late Quaternary sea-level changes and coastal movements in eastern England and eastern Scotland: an assessment of models of coastal evolution. *Quaternary International*, 15/16, 161–173.

Sherman, G. D., Kanetiro, Y. and Matsusaka, Y. 1953: The role of dehydration in the development of laterite. *Pacific Science*, 7, 438–46.

Shi, Z. 1993: Recent saltmarsh accretion and sea level fluctuations in the Dyfi Estuary, central Cardigan Bay, Wales, UK. *Geo-Marine Letters*, 13, 182–8.

Shinn, E. 1973: Sedimentary accretion along the leeward SE coast of the Qatar peninsula, Persian Gulf. In B. H. Purser (ed.), *The Persian Gulf*, Berlin: Springer, 199–209.

Shotton, F. W. 1978: Archaeological inferences from the study of alluvium in the lower Severn-Avon valleys. In S. Limbrey and J. G. Evans (eds), *Man's effect on the landscape: the lowland zone*. Council for British Archaeology Research Report 21, 27–32.

Sieh, K. E. 1978: Slip along the San Andreas Fault associated with the great 1857 earthquake. *Seismology Society of America Bulletin*, 68, 1421–48.

Sieh, K. E. and Jahns, R. H. 1984: Holocene activity on the San Andreas Fault at Wallace Creek, California. *Geological Society American Bulletin*, 95, 883–96.

Sieh, K. E. and Williams, P. L. 1990: Behaviour of the southernmost San Andreas fault during the past 300 years. *Journal of Geophysical Research 95B*, 6629–45.

Sirocko, F., Sarnthein, M., Lange, H. and Erlenkenser, H. 1991: Atmospheric summer circulation and coastal upwelling in the Arabian Sea during the Holocene and the Last Glaciation. *Quaternary Research*, 36, 72–93.

Skaife, S. H. 1955: *Dwellers in darkness*. London: Longmans.

Slattery, M. C. 1990: Barchan migration on the Kuiseb River delta, Namibia. *South African Geographical Journal*, 72, 5–10.

Slaymaker, H. O. 1972: Patterns of present sub-aerial erosion and landforms in mid-Wales. *Transactions of the Institute of British Geographers*, 55, 47–68.

Smalley, I. J. and Krinsley, D. H. 1978: Loess deposits associated with deserts. *Catena*, 5, 53–66.

Smith, R. M., Twiss, P. C., Krauss, R. K. and Bronn, M. J. 1970: Dust deposition in relation to site season, and climatic variables. *Proceedings Soil Science Society of America*, 34, 112–17.

Smith, D. I. and Atkinson, T. C. 1976: Process, landforms and climate in limestone regions. In E. Derbyshire (ed.), *Geomorphology and climate*, London: Wiley, 367–409.

Smith, D. J. 1992: Long-term rates of contemporary solifluction in the Canadian Rocky Mountains. In J. C. Dixon and A. D. Abrahams (eds), *Periglacial geomorphology*, Chichester: Wiley, 203–21.

Smith, R. S. U. 1970: Migration and wind regime of small Barchan dunes within the Algodones dune chain, southwestern Imperial County, California. University of Arizona, Unpublished MSc. thesis.

Smith, D. E., Kolenkiewicz, R., Dunn, P. J. and Torrence, M. H. 1979: The measurement of fault motion by satellite laser ranging. *Tectonophysics*, 52, 59–67.

Sneed, E. D. and Folk, R. L. 1958: Pebbles in the lower Colorado River, Texas, a study in particle morphogenesis. *Journal of Geology*, 66, 114–50.

Soil Science Society of America 1979: Universal soil loss equation: past, present and future. *Soil Science Society of America Special Publication* 8, 1–53.

Sombroek, W. G. 1966: *Amazon Soils*, Wageningen: Centre for Agricultural Publications and Documentation.

Spain, A. V.m Okello-Oloya, T. and Brown, A. J. 1983: Abundances, above-ground masses and basal areas of termite mounds at six locations in tropical north-east Australia. *Revue d'écologie et de biologie du sol*, 20, 547–66.

Spate A. P., Burgess, J. S. and Shevlin J. 1995: Surface lowering by aeolian weathering and salt wedging, Princess Elizabeth Land, Eastern Antarctica. *Earth Surface Processes and Landforms*, in press.

Spencer, T. 1983: Limestone erosion rates and microtopography: Grand Cayman Island, West Indies. Unpublished PhD thesis. University of Cambridge.

Spencer, T. 1985a: Marine erosion rates and coastal morphology of reef limestones on Grand Cayman Island, West Indies. *Coral Reefs*, 4, 59–70.

Spencer, T. 1985b: Weathering rates on a Caribbean reef limestone: results and implications. *Marine Geology*, 69, 195–201.

Spencer, T. 1988: Coastal biogeomorphology. In H. A. Viles (ed.), *Biogeomorphology*. Oxford: Blackwell, 255–318.

Sperling, C. H. B. and Cooke, R. U. 1985: Laboratory simulation of rock weathering by salt crystallisation and hydration processes in hot, arid environments. *Earth Surface Processes and Landforms*, 10, 541–55.

Spitzy, A. and Leenheer, J. 1991: Dissolved organic carbon in rivers. In E. T. Degens, S. Kempe and J. E. Richey (eds), *Biogeochemistry of major world rivers*, Chichester: Wiley. Ch. 9.

Stanley, D. J. and Chen, Z. 1993: Yangtze delta, eastern China: I. Geometry and subsidence of Holocene depocenter. *Marine Geology*, 112, 1–11.

Statham, I, 1981: Slope process. In Goudie, A. S. (ed.), *Geomorphological Techniques*. London: Allen and Unwin, 150–80.

Steen, B. 1979: A possible method for the sampling of Saharan dust. In C. Morales (ed.), *Saharan Dust*. Chichester: Wiley, 279–86.

Stephenson, W. 1961: Experimental studies on the ecology of intertidal environments at Heron Island II: the effects of substratum. *Australian Journal of Marine and Freshwater Research*, 12, 164–76.

Stoddart, D. R. 1990: Coral reefs and islands and predicted sea-level rise. *Progress in Physical Geography*, 14, 521–36.

Strakhov, A. N. 1967: *Principle of lithogenesis* (3 vols). Edinburgh: Oliver and Boyd.

Sugden, D. E. 1978: Glacial erosion by the Laurentide ice sheet. *Journal of Glaciology*, 83, 367–91.

Summerfield, M. A. 1991a: *Global geomorphology*. Harlow: Longman.

Summerfield, M. A. 1991b: Sub-aerial denudation of passive margins: regional elevation versus local relief models. *Earth and Planetary Science Letters*, 102, 460–9.

Summerfield, M. A. and Kirkbride, M. P. 1992: Climate and landscape response. *Nature*, 355, 306.

Sunamura, T. 1983: Processes of sea cliff and platform erosion. In P. D. Komar (ed.), *Handbook of coastal processes and erosion*. Baton Rouge: CRC Press, 233–65.

Sunamura, T. 1992: *Geomorphology of rocky coasts*. Chichester: Wiley.

Svensson, G. 1988: Bog development and environmental condition as shown by the stratigraphy of Store Mosse mire in southern Sweden. *Boras*, 17, 89–111.

Swanston, D. N. and Swanson, F. J. 1976: Timber harvesting, mass erosion and steepland forest geomorphology in the Pacific north-west. In D. R. Coates (ed.), *Geomorphology and Engineering*. Stroudberg: Dowden, Hutchinson and Ross, 199–221.

Sweet, M. L., Neilson, J., Havholm, K. and Farrelley, J. 1988: Algodones dune field of southern California: case history of a migrating modern dune field. *Sedimentology*, 35, 939–52.

Talbot, C. J. and Jarvis, R. J. 1984: Age, budget and dynamics of an active salt extrusion in Iran. *Journal of Structural Geology*, 6, 521–33.

Tetzlaff, G., Peters, M., Janssen, W. and Adams, L. J. 1989: Aeolian dust transport in West Africa. In M. Leinen and M. Sarnthein (eds), *Palaeoclimatology and Palaeometeorology: Modern and Past Patterns of Global Atmospheric Transport*, Kluwer: Dordrecht, 185–203.

Thomas, D. S. G. 1992: Desert dune activity: concepts and significance. *Journal of Arid Environments*, 22, 31–8.

Thomas, M. F. 1992: Ages and relationships of saprolite mantles. Paper presented at BGRG Annual conference, University of Sussex, 26 September 1992.

Thompson, L. G. and Mosley-Thompson, E. 1981: Microparticle concentration variations linked with climatic change: evidence from polar ice cores. *Science*, 212, 812–15.

Thompson, L. G., Mosley-Thompson, E., Davis, J. F., Bolzan, J. F., Dai, J. and Klein, L. 1990: Glacial stage ice-core records from the subtropical Dunde Ice Cap, China. *Annals of Glaciology*, 14, 288–97.

Thornes, J. B. and Brunsden, D. 1977: *Geomorphology and time*. London: Methuen.

Tiller, K. G., Smith, L. H. and Merry, R. H. 1987: Accessions of atmospheric dust east of Adelaide, South Australia and implications for pedogenesis. *Australian Journal of Soil Research*, 25, 43–54.

Tinkler, K. J. 1993: *Field Guide Niagara Peninsula and Niagara Gorge*. Hamilton, Ontario: McMaster University.

Tooley, M. J. 1978: *Sea-level changes: north-west England during the Flandrian stage*. Oxford: Oxford University Press.

Tooley, M. J. 1993: Long-term changes in eustatic sea level. In R. A. Warrick, E. M. Barrow and T. M. L. Wigley (eds), *Climate and sea level change: observations, projections and implications*, Cambridge: Cambridge University Press, 81–107.

Torunski, H. 1979: quoted in Spencer 1988.

Toy, T. J. 1983: A comparison of LEMI and erosion pin techniques. *Zeitschrift für Geomorphologie, Supplementband*, 46, 25–34.

Trendall, A. F. 1962: The formation of 'apparent peneplains' by a process of combined laterisation and surface wash. *Zeitschrift für Geomorphologie*, NF6, 183–97.

Tricart, J. 1972: *The landforms of the humid tropics, forests and savannas.* London: Longman.

Tricker, A. S. and Scott, G. 1980: Spatial patterns of chemical denudation in the Eden catchment, Fife. *Scottish Geographical Magazine*, 96, 114–20.

Trimble, S. W. 1977: The fallacy of stream equilibrium in contemporary denudation studies. *American Journal of Science*, 277, 876–87.

Trimble, S. W. 1988: The impact of organisms on overall erosion rates within catchments in temperate regions. In H. A. Viles (ed.), *Biogeomorphology*, Oxford: Blackwell, 83–142.

Trimble, S. W. and Bube, K. P. 1990: Improved reservoir trap efficiency prediction. *The Environmental Professional*, 12, 255–72.

Trimble, S. W. and Carey, W. P. 1992: A comparison of the Brune and Churchill methods for computing sediment yields applied to a reservoir system. *United States Geological Survey Water Supply Paper 2340*, 195–202.

Trudgill, S. T. 1975: Measurement of erosional weight loss of rock tablets. *BGRG Technical Bulletin*, 17, 13–19.

Trudgill, S. T. 1976: The marine erosion of limestones on Aldabra Atoll, Indian Ocean. *Zeitschrift für Geomorphologie Supplementband*, 26, 164–200.

Trudgill, S. T. 1977: Problem in the estimation of short-term variations in limestone erosion processes. *Earth Surface Processes* 2, 251–6.

Trudgill, S. T., High, C. J. and Hanna, F. K. 1981: Improvements in the micro-erosion meter (MEM). *BGRG Technical Bulletin*, 29, 3–17.

Trudgill, S. 1985: *Limestone geomorphology.* London: Longman.

Trudgill, S. T., Viles, H. A., Cooke, R. U., Inkpen, R. J., Heathwaite, A. L. and Houston, J. 1991: Trends in stone weathering and atmospheric pollution at St Paul's Cathedral, London, 1980–1990. *Atmospheric Environment*, 25A, 2851–3.

Trudgill, S. T., Crabtree, R. W., Ferguson, R. I., Ball, J. and Gent, R. 1994: Ten-year remeasurement of chemical denudation on a Magnesian limestone hillslope. *Earth Surface Processes and Landforms*, 19, 109–14.

Tseo, G. 1990: Reconnaissance of the dynamic characteristics of an active Strezlecki Desert longitudinal dune, southcentral Australia. *Zeitschrift für Geomorphologie* NF34, 19–35.

Tsoar, H. 1974: Desert dune morphology and dynamics, Al-Arish. *Zeitschrift für Geomorphologie Supplementband*, 20, 41–61.

Tsoar, H. 1978: The dynamics of longitudinal dunes. Final Technical Report A-ERO 76-G-172. London: European Research Office, US Army, 171.

Van Heceklon, T. K. 1977: Distant source of 1976 dustfall in Illinois and Pleistocene weather models. *Geology*, 5, 693–5.

Vaudour, J. 1986: Travertins holocènes et pression anthropique. *Méditerranée*, 10, 168–73.

Vice, R. B., Guy, H. P. and Ferguson, G. E. 1969: Sediment movement in an area of suburban highway construction, Scott Run Basin, Fairfax County, Virginia, 1961–64. *United States Geological Survey Water Supply Paper*, 1591-E.

Viles, H. A. 1988: *Biogeomorphology.* Blackwells: Oxford.

Viles, H. A. and Goudie, A. S. 1990: Tufas, travertines and allied carbonate deposits. *Progress in Physical Geography*, 14, 19–41.

Viles, H. and Goudie, A. S. 1992: Weathering of limestone columns from the Weymouth seafront, England. *Proceedings 7th International Congress on Deterioration and Conservation of Stone, Lisbon*, vol. 1, 297–304.

Viles, H. and Trudgill, S. T. 1984: Long-term remeasurements of micro-erosion meter rates, Aldabra Atoll, Indian Ocean. *Earth Surface Processes and Landforms*, 9, 89–94.

Vita-Finzi, C. 1986: *Recent earth movements – an introduction to neotectonics*. London: Academic Press.

Vitek, J. D. and Giardino, J. R. 1987: Rock glaciers: a review of the knowledge base. In J. R. Giardino, J. F. Shroder and J. D. Vitek (eds), *Rock glaciers*, Boston: Allen and Unwin, 1–26.

Voslamber, B. and Veen, A. W. L. 1985: Digging by badgers and rabbits on some wooded slopes in Belgium. *Earth Surface Processes and Landforms*, 10, 79–82.

Vyskocil, P., Wassef, A. M. and Green, R. (eds) 1983: Recent crustal movements, 1982. *Tectonophysics*, 97, 1–351.

Walker, D. 1970: Direction and rate in some British post-glacial hydroseres. In D. Walker and R. G. West (eds), *Studies in the Vegetational History of the British Isles*, Cambridge: Cambridge University Press, 117.

Walling, D. E. and Gregory, K. J. 1970: The measurement of the effects of building construction on drainage basin dynamics. *Journal of Hydrology*, 11, 129–44.

Walling, D. E. and Webb, B. W. 1978: Mapping solute loadings in an area of Devon, England. *Earth Surface Processes*, 3, 85–99.

Walling, D. E. and Kleo, A. H. A. 1979: Sediment yields of rivers in areas of low precipitation: a global view. *International Association of Scientific Hydrology Publication*, 128, 479–93.

Walling, D. E. and Webb, B. W. 1981a: Water quality. In J. Lewin (ed.), *British Rivers*, London: Allen and Unwin, 126–69.

Walling, D. E. and Webb, B. W. 1981b: The reliability of suspended sediment load data. *International Association of Scientific Hydrology Publication*, 133, 177–94.

Walling, D. E. and Webb, B. W. 1983: The dissolved loads of rivers: a global overview. *International Association of Scientific Hydrology Publication*, 141, 3–20.

Walling, D. E. 1984: Dissolved loads and their measurement. In R. F. Hadley and D. E. Walling (eds), *Erosion and sediment yield. Some methods of measurement and modelling*. Norwich: Geobooks, 111–17.

Walling, D. E. and Webb, B. W. 1986: Solutes in river systems. In S. T. Trudgill (ed.), *Solute Processes*, Chichester: John Wiley, 251–327.

Walling, D. 1987: Rainfall, runoff and erosion of the land: a global view. In K. J. Gregory (ed.), *Energetics of Physical Environment*, Chichester: Wiley, 89–117.

Walling, D. E., Quine, T. A. and He, Q. 1992: Investigating contemporary rates of floodplain sedimentation. In P. A. Carling and G. E. Petts (eds), *Lowland Floodplain Rivers: Geomorphological Perspectives*, Chichester: Wiley, 165–184.

Walther, J. 1900: *Das Gesetz der Wüstenbildung in Gegenwart und Vorzeit*. Leipzig: Quelle and Meyer.

Warburton, J. and Beecroft, I. 1993: Use of meltwater stream material loads in the estimation of glacial erosion rates. *Zeitschrift für Geomorphologie*, 37(1), 19–28.

Ward, J. D. 1984: Aspects of the Cenozoic geology in the Kuiseb Valley, central Namib Desert. University of Natal. Unpublished PhD thesis.

Ward, P. R. B. 1984: Measurement of sediment yield. In R. J. Hadley and D. E. Walling (eds), *Erosion and sediment yield: some methods of measurement and modelling*, Norwich: Geo Books, 37–70.

Warren, A. 1988: The dynamics of network dunes in the Wahiba Sands; a progress report. *Journal of Oman Studies Special Report*, 3, 169–81.

Warthin, A. S. 1959: Ironshore on some W. Indian islands. *Transactions New York Academy of Sciences*, 2/21, 649–52.

Watanabe, H. and Ruaysoongnern, S. 1984: Cast production by the megacolecid earthworm *Pheretima sp.* in northeastern Thailand. *Pedobiologia*, 26, 37–44.

Waylen, M. J. 1979: Chemical weathering in a drainage basin underlain by Old Red Sandstone. *Earth Surface Processes*, 4, 167–78.

Weertman, J. 1964: Rate of growth or shrinkage of non equilibrium ice sheets. *Journal of Glaciology*, 5, 145–58.

Weisrock, A. 1986: Variations climatiques et périodes de sédimentation carbonatée à l'Holocène – l'âge des dépôts. *Méditerranée*, 10, 165–7.

Wellman, P. 1988: Tectonic and denudational uplift of Australian and Antarctic highlands. *Zeitschrift für Geomorphologie*, 32, 17–29.

Wessels, D. C. J. and Schoemann, B. 1988: Mechanism and rate of weathering of Clarens Sandstone by an endolithic lichen. *South African Journal of Science*, 84, 274–77.

Whalley, W. B. 1983: Desert varnish. In A. S. Goudie and K. Pye (eds), *Chemical sediments and geomorphology*. London: Academic Press, 197–226.

White, W. A. 1972: Deep erosion by continental ice sheets. *Bulletin Geological Society of America*, 83, 1037–56.

Whitehouse, I. E., Knuepfer, P. L. K., McSaveney, M. J. and Chinn, T. J. H. 1986: Growth of weathering rinds on Torlesse Sandstone, southern Alps, New Zealand. In S. M. Colman and D. P. Dethier (eds), *Rates of chemical weathering of rocks and minerals*. Orlando: Academic Press, 419–35.

Williams, C. and Yaalon, D. H. 1977: An experimental investigation of reddening in dune sand. *Geoderma*, 17, 181–91.

Williams, M. A. J. 1968: Termites and soil development near Brocks Creek, Northern Territory. *Australian Journal of Science*, 31, 153–4.

Williams, P. W. 1982: Speleothem dates, Quaternary terraces and uplift rates in New Zealand. *Nature*, 298, 257–9.

Williams, S. H. 1981: Calculated and inferred aeolian abrasion rates: Earth and Mars. MS thesis, Arizona State University, Tempe.

Wilson, I. G. 1971: Desert sandflow basins and a model for the development of ergs. *Geographical Journal*, 137, 180–99.

Wilson, L. 1973: Variations in mean annual sediment yield as a function of mean annual precipitation. *American Journal of Science*, 273, 335–49.

Wilson, M. E. and Wood, S. H. 1980: Tectonic tilt rates derived from lake-level measurements, Salton Sea, California. *Science* 207, 183–6.

Winkler, E. M. 1975: *Stone: properties, durability in man's environment* (2nd edn). Vienna: Springer.

Wischmeier, W. H. and Smith, D. D. 1978: Predicting rainfall erosion losses. *USDA Agriculture Research Handbook*, 537.

Wise, S. 1980: Cs-137 and Pb-210: review of the techniques and some applications in geomorphology. In J. Lewin, R. Cullingford and D.

Davidson (eds), *Timescales in Geomorphology*. New York: Wiley, 109–27.

Wolman, M. G. 1967: A cycle of sedimentation and erosion in urban river channels. *Geografiska Annaler*, 49A, 385–95.

Wolman, M. G. and Miller, J. P. 1960: Magnitude and frequency of forces in geomorphic processes. *Journal of Geology*, 68, 54–74.

Wolman, M. G. and Schick, A. P. 1967: Effects of construction on fluvial sediment, urban and suburban areas of Maryland. *Water Resources Research 3*, 451–64.

Wood, T. G. and Sands, W. A. 1978: The role of termites in ecosystems. In M. V. Brian (ed.), *Production ecology of ants and termites*, Cambridge: Cambridge University Press, 245–92.

Woodroffe, C. D. 1990: The impact of sea level rise on mangrove shorelines. *Progress in Physical Geography*, 483–520.

Woodruff, N. P. and Siddoway, F. H. 1965: A wind erosion equation *Proceedings of the Soil Science Society of America*, 29, 602–8.

Wright, J. and Smith, B. 1993: Fluvial comminution and the production of loess-sized quartz silt. A simulation study. *Geografiska Annaler*, 75(A), 25–34.

Yaalon, D. H. and Ganor, E. 1975: Rate of aeolian dust accretion in the Mediterranean and desert fringe environments of Israel. *19th International Congress of Sedimentology*, theme 2, 169–74.

Yair, A. and Enzel, Y. 1987: The relationship between annual rainfall and sediment yield in arid and semi-arid areas. The case of the northern Negev. *Catena Supplement*, 10, 121–35.

Yair, A. and Rutin, J. 1981: Some aspects of regional variation in the amount of available sediment produced by isopods and porcupines, northern Negev, Israel. *Earth Surface Processes and Landforms*, 6, 221–34.

Young, A. 1960: Soil movement by denudational processes on slopes. *Nature*, 188, 120–2.

Young, A. and Saunders, I. 1986: Rates of surface processes and denudation. In A. D. Abrahams (ed.), *Hillslope Processes*. Boston: Allen and Unwin, 1–27.

Young, R. and McDougall, I. 1993: Long-term landscape evolution: early Miocene and modern rivers in southern New South Wales, Australia. *Journal of Geology*, 101, 35–49.

Zenkovich, V. P. 1967: *Processes of coastal development*. Edinburgh: Oliver and Boyd.

Zeitler, P. K., Johnson, N. H., Naeser, C. W. and Tahirkheli, R. A. K. 1982: Fission track evidence for Quaternary uplift of the Nanga Parbat Region, Pakistan. *Nature*, 298, 255–7.

Index

Note: Page references to figures are in *italic*; page references to tables are in **bold**; P after page reference indicates a plate.